ADVANCES IN
ELECTRON TRANSFER
CHEMISTRY

Volume 1 • 1991

ADVANCES IN ELECTRON TRANSFER CHEMISTRY

A Research Annual

Editor: PATRICK S. MARIANO
Department of Chemistry and Biochemistry
University of Maryland-College Park

VOLUME 1 • 1991

Greenwich, Connecticut *London, England*

55 Old Post Road, No. 2
Greenwich, Connecticut 06830

JAI PRESS LTD.
118 Pentonville Road
London N1 9JN
England

ISBN: 1-55938-167-1

Manufactured in the United States of America

CONTENTS

LIST OF CONTRIBUTORS vii

INTRODUCTION TO THE SERIES:
AN EDITOR'S FOREWORD
Albert Padwa ix

PREFACE
Patrick S. Mariano xi

PHOTOINDUCED ELECTRON TRANSFER ON
IRRADIATED SEMICONDUCTOR SURFACES
Marye Anne Fox 1

THERMAL AND PHOTOCHEMICAL ACTIVATION OF
AROMATIC DONORS BY ELECTRON TRANSFER
Christian Amatore and Jay K. Kochi 55

DISTANCE AND ANGLE EFFECTS ON ELECTRON
TRANSFER RATES IN CHEMISTRY AND BIOLOGY
George L. McLendon and Anna Helms 149

ELECTRON TRANSFER REACTIONS FOLLOWED BY
RAPID BOND CLEAVAGE: INTRA-ION PAIR ELECTRON
TRANSFER OF PHOTOEXCITED CYANINE BORATES
AND CHEMICALLY INITIATED ELECTRON-
EXCHANGE LUMINESCENCE
Gary B. Schuster 163

LIST OF CONTRIBUTORS

Christian Amatore
Laboratoire de Chimie
Ecole Normale Superieure
Paris, France

Marye Anne Fox
Department of Chemistry
University of Texas at Austin
Austin, Texas

Anna Helms
Department of Chemistry
University of Rochester
Rochester, New York

Jay K. Kochi
Department of Chemistry
University of Houston
Houston, Texas

George L. McLendon
Department of Chemistry
University of Rochester
Rochester, New York

Gary B. Schuster
Department of Chemistry
University of Illinois-
Urbana-Champaign
Urbana, Illinois

INTRODUCTION TO THE SERIES: AN EDITOR'S FOREWORD

The field of organic chemistry has developed dramatically during the past forty years. Thus it appears to be an opportune time to publish a series of essays on various relevant themes in the 1990s written by workers who are active in the discipline. This collection includes many of the important areas of current research interest. To cover such a broad area a very substantial effort was needed, as was the cooperation of a large number of colleagues and friends who have agreed to act as series editors. I have been gratified by the favorable response of research workers in the field to the invitation to contribute chapters in their own specialties. Contributors have written critical, lively, and up-to-date descriptions of their field of interest and competence, so that the chapters are not merely literature surveys. It is hoped that this new and continuing series will prove valuable to active researchers and that many ideas will be generated for future theoretical and experimental research. The wide coverage of material should be of interest to graduate students, postdoctoral fellows, and those teaching specialized topics to graduate students.

Department of Chemistry
Emory University
Atlanta, Georgia

Albert Padwa
Consulting Editor

PREFACE

The consideration of reaction mechanisms involving the movement of single electrons is relatively new in the fields of chemistry and biochemistry. Despite this, studies conducted in recent years have uncovered a large number of chemical and enzymatic processes that proceed via single electron transfer pathways. At the current time numerous investigations are underway probing the operation of electron transfer reactions in organic, organometallic, biochemical, and excited state systems. In addition, theoretical and experimental studies are being conducted to gain information about the factors that govern the rates of single electron transfer. It is clear from the current level of activity that electron transfer chemistry has now become one of the most active areas of chemical study.

The series, *Advances in Electron Transfer Chemistry*, has been designed to allow scientists who are developing new knowledge in this rapidly expanding area to describe their most recent research findings. Each contribution will be in a minireview format focusing on the individual author's own work as well as the studies of others that address related problems. Hopefully, by following this protocol, *Advances in Electron Transfer Chemistry* will serve as a useful series for those interested in learning about current breakthroughs in this rapidly expanding area of chemical research.

Patrick S. Mariano
Series Editor

PHOTOINDUCED ELECTRON TRANSFER ON IRRADIATED SEMICONDUCTOR SURFACES

Marye Anne Fox

1. Introduction 2
2. Controlling Photoelectrochemical Transformations 7
   2.1. Adsorption Effects in Photoelectrochemistry 7
   2.2. Potential Control 11
   2.3. Current Control 14
3. Organic Reactions on Irradiated Semiconductors 19
   3.1. Solvent 20
   3.2. Photooxidation/Photodehydrogenation 22
   3.3. Organic Photoelectrochemical Oxidations 23
   3.4. Organic Photoelectrochemical Reductions 38
   3.5. Photoelectrochemical Transformations with Reversible Electron Transfer as the Activation Step 39
4. Opportunities and Questions in Photoelectrochemistry 40
   4.1. Sensitization 41
   4.2. New Semiconductors and Supports 41
   4.3. Applications 43
Acknowledgments 43
References 44

Advances in Electron Transfer Chemistry,
Volume 1, pages 1–53.

ISBN: 1-55938-167-1

1. INTRODUCTION

By definition, a semiconductor possesses band structure. A filled band of closely spaced molecular orbitals is separated from a vacant (or nearly vacant) band of similarly closely space orbitals. Since the filled band relates to bonding in the component material, the filled band is called the valence band and electrons occupying this band are highly localized between atoms. Electrons placed within the vacant band, in contrast, are highly delocalized and therefore contribute significantly to conduction in the material. This energetically higher lying orbital is thus called the conduction band. In a good conductor, i.e., a metal, the valence and conduction bands overlap to form a near continuum of states. In contrast, in an insulator, a large gap between the valence and conduction bands occurs, making electronic interaction between these bands exceedingly difficult. The structure of a semiconductor lies between these extremes: a gap exists between the valence and conduction bands, the width of which allows for some electronic equilibration in the ground state and which permits photoexcitation in the visible- to long-wavelength ultraviolet ranges.

Photoexcitation of a semiconductor thus promotes an electron from the valence band to the conduction band. The electron deficiency in the valence band is called a hole, and, like electrons in the valence band, the hole is much less mobile than the electron in the conduction band. Thus, formation of a photogenerated electron-hole pair achieves not only highly active oxidizing (hole) and reducing (electron) equivalents, but also spatial separation between possible redox sites.

Most photoelectrochemical studies of semiconductors have involved these solid materials immersed in a liquid electrolyte. Although the characteristics of the semiconductor–liquid junction under illumination have been extensively reviewed,[1–14] its salient features can be briefly summarized as in Figures 1 and 2. An intrinsic semiconductor can be represented as possessing a Fermi level equidistant between the valence and conduction bands (Figure 1). On doping with an electron donor, a slight excess of electrons is deposited in the conduction band, shifting the Fermi level to a position just below the conduction band edge in this negatively doped (n-type) semiconductor. If an electron-accepting dopant had been introduced instead, the electronically depleted Fermi level would lie near the top of the valence band edge in this positively doped (p-type) material.

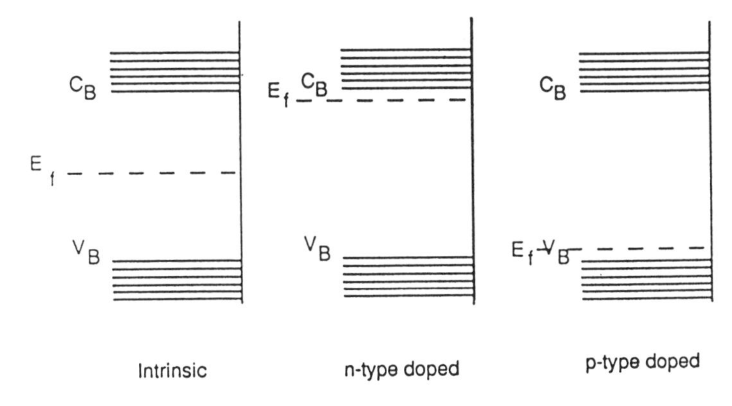

Figure 1. Shift of the Fermi Level with Doping.

Before immersing a semiconductor in an electrolyte containing a poised redox couple, energy levels (with respect either to the absolute vacuum or to a reference electrode, which is experimentally easier) can be stipulated. These redox levels are physical properties of the respective semiconductor and redox couple. The levels of an arbitrarily chosen system are shown in Figure 2 before and after immersion. Bringing the semiconductor into contact with the redox couple causes the Fermi level in the bulk of the solid to equilibrate with the solution phase redox couple. This causes the development of an electrical field at the interface, where a space–charge layer forms. In an n-type semiconductor, this causes both

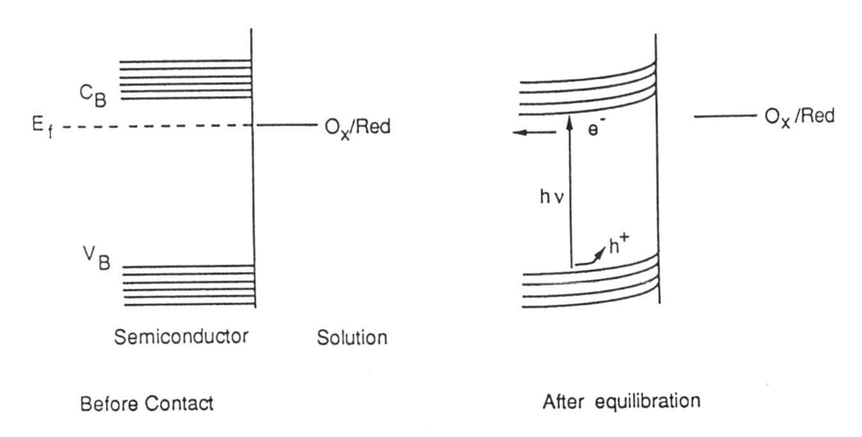

Figure 2. Equilibration of an n-Type Semiconductor with a Solution Phase Redox Couple.

the valence and conduction band to bend upward as one moves from the bulk to the interface. Photoexcitation partially repopulates the conduction band, causing this band bending to partially diminish, as light perturbs the system from equilibrium. Band bending also ensures that photochemical excitation within the space–charge layer will cause electrostatic directional movement of the photogenerated electron away from the irradiated interface into the bulk (as thermal equilibrium is restored), while a valence band hole migrates toward the interface. The surface-bound hole is then trapped by interfacial electron transfer from a solution-phase donor, while the electron migrates from the bulk through an external circuit to the counterelectrode where it effects a reductive half-reaction. Light thus provides the driving force for oxidation on the surface of the n-type semiconductor and reduction at a connecting counterelectrode (often a metal).

On a p-type semiconductor, the direction of the band bending is reversed. The resulting space–charge layer will thus display the opposite sign, and illumination will cause interfacial trapping by an electron acceptor (i.e., semiconductor-mediated reductions) with oxidation at the counterelectrode supplying the lost electron to the bulk semiconductor. Although equally attractive in principle, photocatalytic reductions on p-type semiconductors have been less extensively studied than photocatalytic oxidations on n-type semiconductors because the p-type semiconductors are far less stable under operating conditions than the analogous n-type materials. Appreciable corrosion of the semiconductor surface thus makes difficult their practical utilization in effecting photocatalytic redox reactions.

The photoelectrochemical cell summarized succinctly in Figure 2 can be further simplified, as was first recognized by Bard,[3,7,9] if the counterelectrode is short-circuited onto the semiconductor. The resulting metallized semiconductor particle shown in Figure 3 contains all the components of a full electrochemical cell: a photoactive semiconductor surface, a metal counterelectrode, and a conductive pathway (the con-

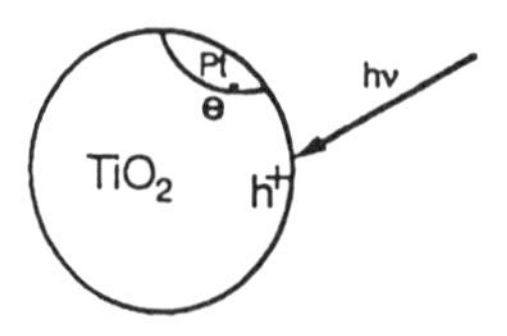

Figure 3. A Semiconductor Powder as a Photodiode.[3]

duction band) that allows for electrical interconnection and acts as a photochemical diode. Irradiation creates an electron-hole pair, which in an n-type semiconductor particle will initiate oxidation of a donor adsorbed on the semiconductor surface and reduction of an acceptor adsorbed on the metal island counterelectrode. An electrolyte is no longer necessary if the respective donors and acceptors are effectively adsorbed, since the transfer of charge through solution need not take place. Furthermore, these characterizations will apply to a variety of particle sizes: interfacial electron transfer can be accomplished on single crystals, polycrystalline films, powders, colloids, or, in at least some cases, even small clusters of semiconductors, with or without external supports. This cell can be run simply as a suspension in a solution containing the desired redox couple, making for a facile experimental setup but complicating the optics involved, for example, in quantum yield determinations.

Even this suspension can be simplified, for metallization of semiconductor particle (to deposit a counterelectrode) will be necessary only for systems in which the oxidative and reductive half-reactions are either slow or kinetically indistinguishable. That is, if interfacial hole trapping is rapid (compared with the rate of electron-hole recombination), the photogenerated electron will be confined to the conduction band. It can then migrate, relatively leisurely, through the particle until it finds a defect or other surface site where interfacial reduction can occur. Thus, for many photocatalytic redox reactions, surface metallization of the semiconductor surface will not be necessary.

One significant exception to this generalization (and a reason for the extensive published work on deposition of metals on semiconductor surfaces) is encountered when multiple electron transfers are required or when gases evolve. In water splitting, for example, a net four electron exchange is needed and the adsorption–desorption equilibrium of two gases (oxygen and hydrogen) from the catalyst surface must be effectively controlled. Similar considerations also apply in the photocatalytic reduction of carbon dioxide and molecular nitrogen.

These studies led to the following oversimplified mechanism for photocatalytic redox reactions occurring on irradiated semiconductor surfaces:

$$\text{semiconductor} \xrightarrow{\text{band gap } h\nu} e^-/h^+ \tag{1}$$

$$h^+ + D_{\text{ads}} \longrightarrow D^+ \tag{2}$$

$$e^- + A_{ads} \longrightarrow A^- \quad (3)$$

$$D^+ \longrightarrow \text{products} \quad (4)$$

$$A^- \longrightarrow \text{products} \quad (5)$$

$$D^+ + A^- \longrightarrow D\text{–}A \quad (6)$$

$$D^+ + O_2 \longrightarrow D - O\text{–}O^+ \quad (7)$$

Photoexcitation generates an electron-hole pair (Eq. 1). Hole trapping by an adsorbed donor results in single electron oxidation to form an adsorbed cation radical (Eq. 2) while electron trapping by an adsorbed acceptor forms the bound anion radical (Eq. 3). Any of a variety of ion radical reactions can consume D^+ and A^- (Eqs. 4 and 5); among these are deprotonations, bond cleavages, and dimerizations. Alternatively, the oppositely charged ion radicals can collapse to form a zwitterion or neutral adduct (Eq. 6). The ion radicals can also be trapped by other adsorbed neutral reagents, as shown in Eq. 7 for trapping of the cation radical by adsorbed oxygen. Clearly the fate of these adsorbed intermediates will depend on the structure of the singly oxidized or reduced species. The reversibility of any of these reactions would obviate product formation and would waste the incident light energy, thus reducing the quantum yield. The competition between the rates for back reaction (i.e., charge recombination) and the dark reaction of the adsorbed radicals will determine the outcome of such photocatalytic conversions.

The highly oxidizing band edge of typical photoactive semiconductors makes for easy oxidation of most substrates bearing either a conjugated π system or nonbonded electrons. The absolute reducing power of a conduction band electron is more modest, however, and most organic substrates lacking either positive charge or heavy substitution with electron-withdrawing groups are reduced at more negative potentials than the conduction band edge makes accessible. Fortunately, the conduction band edge of many metal oxide semiconductors lies almost isoenergetic with the reduction potential of oxygen, making electron trapping by oxygen a reasonable pathway. The facility of the reaction is thus governed by the relative reactivity of the adsorbed ion radical intermediates formed via interfacial electron transfer, trapping the photogenerated electron-hole pair.

These principles are easily transferrable to studies of the mechanisms of photocatalyzed organic transformations occurring on irradiated semiconductor surfaces, an area in which my research group has worked intensely for the last several years. It is the intent of this article to summarize these investigations and to put them in perspective with other investigations of photoelectrochemical phenomena. As such, this article cannot be construed either as a comprehensive survey of either all questions of scientific interest involving photocatalytic redox reactions on semiconductor surfaces or as a complete bibliography on the topics that are addressed here. Examples illustrating a given subject are sometimes randomly chosen and apologies are offered to authors whose work is not cited, even when dealing with similar systems or scientific problems.

2. CONTROLLING PHOTOELECTROCHEMICAL TRANSFORMATIONS

Photocatalytic reactions on illuminated semiconductor surfaces are closely related to those occurring in conventional electrochemical cells. Both photoelectrochemical and electrochemical transformations involve nonhomogeneous arrays in which mass transfer to the solid–liquid junction and adsorption phenomena will control access to an electroactive reagent. Surface modification of either the metal electrode or the photosensitive semiconductor particle can influence both the kinetics for interfacial electron transfer and the steady-state adsorption equilibria of adsorbates. Both types of cells achieve selectivity by controlling the passage of current: with a potentiostat in a conventional electrochemical cell and with light flux in a photoelectrochemical cell. Both attain reaction control by governing accessible potential: again, by use of a potentiostat in a conventional cell, but by the judicious choice of the light-responsive semiconductor (and hence, the band edge positions) and its environment (e.g., pH of a contacting aqueous solution) in photoelectrochemistry. In this section we consider examples of each of these methods for controlling photoelectrochemical reactivity.

2.1. Adsorption Effects in Photoelectrochemistry

Chemical processes involving heterogeneous catalysis will typically involve the following stages:[15] (1) mass transfer of reactants from bulk solution to the external surface of the catalyst (outer coordination or solvation sphere); (2) mass transfer of the reactants to the active site of the catalyst (inner coordination or solvation sphere); (3) adsorption of the reactants at the active site; (4) the surface reaction; (5) desorption of the products from the active site; (6) mass transfer to the external surface of the catalyst; and (7) mass transfer of the product to the bulk of the solution. Although the mean lifetime of an electron-hole pair will depend on the size of a particle (or, more specifically, on the ratio of its surface area to the thickness of the space–charge layer) and on the number of surface states, the photogenerated redox pair typically lives only a fraction of a nanosecond in the absence of appropriate electron or hole traps. This short lifetime precludes activation of any molecules that are not already effectively adsorbed before the absorption of the photon and, hence, the creation of the electron-hole pair. Thus, steps (1), (2), (6), and (7) will not be of concern for understanding the dynamics of photoelectrochemical activation and will affect the steady-state yields of product only in so far as they influence the adsorption–desorption equilibrium, steps (3) and (5).

Kinetic studies have shown that the rates of reactions requiring preadsorption are directly proportional to the surface coverage.[16,17] The Langmuir expression (Eq. 8) describes

$$\Theta = KC/(1 + KC) \qquad (8)$$

the surface coverage Θ as a function of the concentration of the adsorbate C and the equilibrium constant K. This form, which predicts the reciprocal of the reaction rate to depend linearly on the reciprocal of the concentration of the reactant, has been useful in describing the initial stages of many of the liquid phase organic oxidation reactions we will consider in the next section.[16] If this linearity is maintained, but with a variable slope that depends on the concentrations of another reagents, a modified Langmuir isotherm involving preadsorption at different sites is indicated. The gas phase oxidation of ammonia on TiO_2, for example, proceeds via this Langmuir–Hinshelwood mechanism.[18] An alternative kinetic expression, in which the reciprocal of the reaction rate is proportional to the reciprocal square root of the reactant concentration, describes reactions that proceed via dissociative adsorption of the reactant. This expression applies, for example, to the gas phase photocatalytic dehydrogenation of aliphatic alcohols.[19]

Such expressions not only describe the requisite preequilibration but also define the degree of inhibition by reagents in competition for active adsorption sites and the nature of the active adsorption site. Overinterpretation is nonetheless easy: although organic compounds do not compete with adsorbed oxygen for electron capture, the above analysis assumes the same adsorption site for all reactants.

If such limitations are kept in mind, however, Langmuir isotherms can be used to describe relative reactivity in a series of structurally similar reagents. For example, the relative reactivity of a series of isomeric picolines (methyl pyridines) showed the effect of steric hindrance to adsorption in their relative reactivity order. The *para*-substituted isomer was oxidized most readily photoelectrochemically by virtue of its strong adsorption and electronic structure. The less electron-rich *meta*-isomer was more reactive, however, than the *ortho*-isomer, which, despite its electronic similarity to the *para*-isomer, was much less strongly bound to the photocatalyst surface.[20] Similarly, the 7-fold kinetic enhancement for the photoelectrochemical oxidation of 1-pentanol over 2-pentanol was ascribed to the more facile adsorption at an active surface site by the sterically less demanding primary alcohol.[21] Similar effects have also been reported by Pattenden and co-workers who found that synthetically useful yields could be attained in the photoelectrochemical oxidation of primary alcohols to aldehydes, but only much poorer conversions could be realized under the same experimental conditions in the oxidation of secondary alcohols to ketones.[22]

A more subtle example of preferential adsorption effects on photoelectrochemical selectivity can be seen in the regiopreference for monodecarboxylation of remotely substituted cyclohexane dicarboxylic acids.[23] In cyclohexane-1,2-dicarboxylic acid, photocatalytic oxidation induces monodecarboxylation, producing a radical that is either reduced or trapped by adsorbed oxygen (Eq. 9).

$$\text{C}_6\text{H}_{10}(\text{CO}_2\text{H})_2 \xrightarrow{TiO_2^{\bullet}} \text{C}_6\text{H}_{11}\text{CO}_2\text{H} + \text{C}_6\text{H}_{10}\text{=O} + CO_2 \quad (9)$$

The ratio of monoacid to ketone can be adjusted by controlling the oxygen content of the purging gas. The isolation of the monoacid in high yield indicates that the diacid reacts substantially faster than the monoacid product. Presumably the faster reaction of the more strongly adsorbed diacid derives from its better accessibility to surface-confined holes. In the same reaction on a remotely substituted analogue, the regiopreference for formation of *trans*-4-phenylcyclohexane carboxylic acid over the 3-phenyl isomer in the photoelectrochemical oxidation of *trans*-4-phenyl-cyclohexane-*cis*-1,2-dicarboxylic acid (Eq. 10) can be

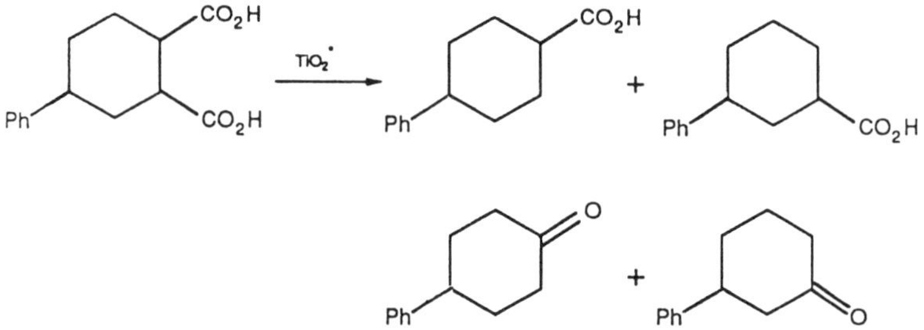

(10)

attributed to reactivity from each of the two possible adsorbed conformations: in the preferred structure, the two carboxylic acids adsorb to the photoactive surface; in the alternate structure, the ring is π-bound, with the 2-carboxylate also adsorbed. The magnitude of these relative adsorption effects also depends on the presence of coadsorbates, i.e., the extent of coplatinization.

Adsorption effects are also sensitive to the photolysis itself. It is clear, for example, that the adsorption of gaseous oxygen can be increased on metal oxide semiconductors by band gap illumination. This effect derives from the relative increase in the number of minority carriers generated in the creation of the electron-hole pair and hence in the development of high localized partial charge.[16] Depending on the identity of the minority carrier, photodesorption effects can also be observed, sometimes upon irradiation with sub-band gap wavelengths. Adsorption desorption equilibria result in isotopic scrambling of surface oxide groups, an effect that can complicate mechanistic studies of photoinduced oxygenation.[16]

Radicals formed via photoinduced interfacial electron transfer appear to also adhere strongly to the surfaces of irradiated semiconductor particles.[24] For example, in the photoinduced decomposition of CF_3C-HBrCl on platinized TiO_2 in methanol, electron attachment proceeds to release F^- by attachment of a second electron to the initially formed $CF_3CHCl^•$ radical. This reactive radical therefore must have remained adsorbed (and protected from other dark reactions) long enough for the second reduction to occur. Similar reduction of an adsorbed radical (to the exclusion of normal free radical reactivity) had also been invoked to explain the formation of monodecarboxylation product in Eq. 4[23] and may be involved in the formation of methane in the simple photo-Kolbe decarboxylation of acetic acid.[25,26] The rapid photooxidation of sulfite on α-Fe_2O_3 similarly proceeds without the behavior typical of freely diffusing $SO_3^{\bullet-}$ radicals.[24] Thus, adsorption effects can not only control the relative reactivity of starting materials, but can also influence the course of radical reactions initiated by the photocatalyst.

2.2. Potential Control

Unlike a metal electrode that can be continuously tuned by an external potentiostat, the accessible potentials relevant to photoelectrochemical oxidations and reductions relate to the band edges of the semiconductor chosen. The band positions of several common semiconductors in contact with aqueous electrolytes are shown in Figure 4.[1,2,14] Of the several inorganic redox couples shown at the far right of the figure, we focus as an example on the proton/hydrogen gas couple. To induce hydrogen evolution from an acidic solution (at pH 1), one must choose a semiconductor whose conduction band edge lies negative of the desired reduction potential. Thus, a photogenerated electron in the conduction band of TiO_2 is sufficiently energetic to generate molecular hydrogen if an appropriate gas evolution catalyst is present, but SnO_2 is not. One of the goals in the synthesis of new semiconductors is to prepare materials that resist corrosion yet possess substantially more negative conduction band edges, which would permit wider photocatalytic reduction activity. Similar band edge considerations also apply to oxidizability and the valence band position. Clearly, the metal oxides possess valence band edges lying at highly positive potentials, which affirms their high reactivity as photocatalysts for products derived from single electron oxidation.

These band edges, which are usually characterized as flat band potentials from capacitance-potential data with a Mott–Schottky analysis[27] or from photocurrent measurements,[28] shift with pH and solvent. The bands of TiO_2, for example, shift negative by about 59 mV for each pH unit or about 0.7 V when measured in acetonitrile (neutral) compared with those in water at pH 1.[29] The measured capacity of a semiconductor–electrolyte interface is also related to the space–charge capacitance of the interface and to the dopant concentration. Such band edge measurements thus are helpful not only in a thermodynamically predictive sense, but also as a vehicle for characterizing new semiconductors and determining the presence of bulk localized states. That similar characterizations can be made for integrated semiconductor systems[30] is important for the use of powders, colloids, and quantum sized particles as photocatalysts.

The cathodic shift observed in moving from water to photoelectrochemically inert solvent is very important for photocatalytic oxygenation reactions, for it allows oxygen to act as an electron trap since its reduction potential becomes nearly isoenergetic with the conduction band edges of the common metal oxides. (The reduction potential of O_2 in acetonitrile is −0.78 V.[31]) Thus, TiO_2 can act as a photocatalyst for the activation of oxygen.

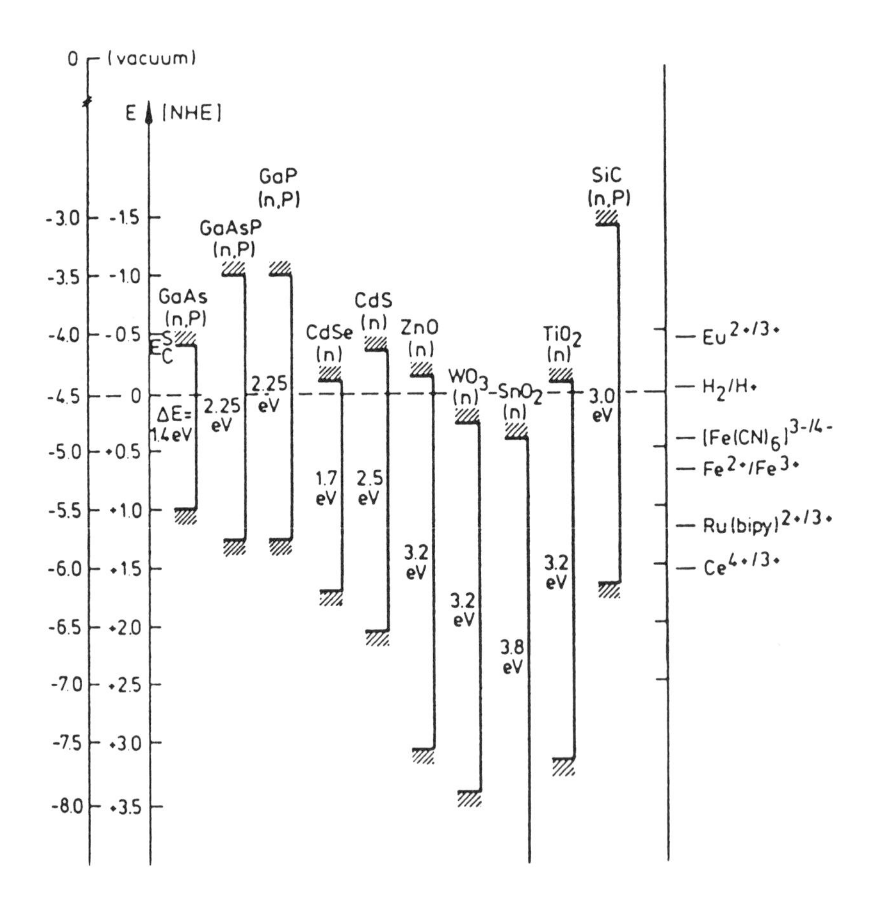

Figure 4. Band Edge Positions of Several Semiconductors in Contact with Aqueous Electrolyte at pH 1.[14]

If the valence band edges dictate thermodynamic limitations for photoinduced interfacial electron transfer, judicious choice of the semiconductor should allow for chemoselectivity for oxidation reactions occurring on illuminated semiconductor suspensions. For example, the photogenerated hole in TiO_2 is much more highly oxidizing than one in CdS (Figure 4). In accord with this expected order of reactivity, organic carboxylic acids can be much more efficiently decarboxylated on band gap irradiation on TiO_2 than on CdS.[32]

Likewise, divergent products are obtained in the photoelectrochemical oxidation of lactic acid on these two photocatalysts.[33] Platinized TiO_2 gave decarboxylation, whereas CdS caused oxidation of the alcohol group, giving pyruvic acid as the major product. The obvious interpretation of valence band edge chemoselectivity was clouded, however, by the observation that potential-controlled oxidation of lactic acid of a metal electrode at +1.3, +1.5, and +1.6 V (respectively, potentials positive of, approximately equivalent to, and negative of the valence band edge of CdS) led to decarboxylation at each potential. Furthermore, nearly equivalent current doubling was observed on CdS and on TiO_2, suggesting a close similarity in mechanism on the two photocatalysts. It is possible therefore that what originally appeared to be potential control may in fact have been caused instead by adsorption differences on the two different surfaces. A recent reinvestigation also revealed a remarkable pH dependence in which pyruvic acid becomes the main product on TiO_2 at pH greater than 12. The suggested interpretation postulated direct oxidation of lactic acid (with decarboxylation) at low pH on Pt, with the oxidation of water (or OH^-) becoming dominant at high pH. Hydrogen abstraction by the resulting hydroxyl radical would then produce a ketyl radical, diproportionation of which would form pyruvic acid and regenerate starting material. The feasibility of this pathway was supported by similar reactivity with Fenton's reagent.

de Mayo's group has reported several examples in which differential reactivity of similar substrates is attributed to the relative levels of the substrate's oxidation potential and the illuminated semiconductor's band edge. For example, the geometric isomerization of diarylcyclopropanes (via ring opening of the singly oxidized cation radical intermediate) occurred on CdS only with aryl groups substituted with strong electron-releasing groups so as to place the oxidation level of the hydrocarbon above the CdS valence band edge[34a] and only those strained hydrocarbons whose oxidation potential lay more positive than the CdS valence band edge could be isomerized to the more stable ring-opened isomer.[34b]

Kisch has also attributed the lower quantum efficiency of CdS than ZnS in the photocatalytic dehydrodimerization of dihydrofurans to the less favorable band edge position of CdS and the shorter lifetime of the electron-hole pair in that photocatalyst.[35] In a recent report on the irradiated semiconductor-mediated cleavage of DNA, the relative activity of several metal oxides was found to correlate with the energy of the valence band edge.[36] In none of these cases, however, were the absolute band positions assigned nor were adsorption differences that had proved to be critical in the selectivity observed with lactic acid clearly sorted.

Thus, although theory implies that chemoselectivity from mixtures, or regioselectivity within a multifunctional molecule, should be attainable via variation of semiconductor band edges, such conclusions are difficult to unambiguously establish. Nor has a well-defined case of possible kinetic retardation in a highly exothermic interfacial electron transfer to a photogenerated hole in a very positive, highly oxidizing valence band (an effect parallel to the Marcus inverted region in homogeneous electron transfer reactions) been unambiguously documented. Recent results in our laboratory indicate, however, that the kinetic competition between amines and alcohols for photoinitiated oxidative cleavage may be controlled by such factors.[37]

2.3. Current Control

2.3.1. Number of Electrons in a Redox Transformation

One significant advantage of a photoelectrochemical cell over a conventional electrochemical cell is that the highly oxidizing and reducing equivalents of the electron-hole pair can be made available on the timescale required for photoexcitation, with the local current density being proportional to the incident light intensity. The ability to adjust light flux can thus permit control of the number of electrons passed across the interface: if the number of surface-accessible electron-hole pairs is low compared with the number of adsorbed reactant molecules, the high redox activity of the electron-hole pair is extinguished by interfacial electron transfer. That is, at low light flux, transfer of an electron from an adsorbed donor fills the photogenerated hole, causing the effective surface potential to revert to the quasi-Fermi level. Similar reduction effects should also be realized with adsorbed electron traps.

This control makes it possible to trap redox transients (often radicals) at intermediate oxidation levels. In the Kolbe electrolysis, for example, hole trapping by an adsorbed carboxylic acid or carboxylate generates a carboxylate radical whose oxidation potential may be less positive than the starting material. In such cases, two electron oxidation to generate a carbocation may occur in competition with radical dimerization. In the photo-Kolbe reaction, however, the first oxidation effects hole trapping. rendering the surface of the irradiated semiconductor reductive rather than oxidative. The radical thus remains bound to the surface, ultimately being reduced to the observed alkane (Eq. 11).[25,38]

$$CH_3CO_2H \xrightarrow{TiO_2^{\bullet}} CH_4 + CO_2 \qquad (11)$$

Likewise, the electrochemical oxidation of vicinal diacids is known to proceed through an analogous two electron oxidation, generating an alkene.[39] This electrochemical deprotection involves the formation of a very easily oxidized β-carboxylate radical, which, at the potential necessary to effect the oxidation of the first carboxylic acid, is oxidized to a cation, deprotonation and carboxylation of which produce the freed double bond. In the photoelectrochemical route, however, interfacial trapping of the photogenerated hole produces a surface-bound, stabilized radical, which is reductively trapped as the protonated anion, producing thereby the observed monodecarboxylation product.[23]

If the ratio of incident photons to adsorbed molecules is high, however, the semiconductor can act as an electron pool, making multiple electron transfer processes possible. On illumination, semiconductor particles can become charged, permitting observable electrophoretic mobility even of large diameter particles.[40] The build up of charge in colloidal metal oxides is obvious from the absorption spectrum of the electron observed in acidic TiO_2 sols, which can be recognized easily by the intense blue color in the absence of air.[41,42] EPR experiments have shown that these trapped electrons are found at Ti^{4+} ions on the particle surface.[42] Time resolution of the dynamics for charge carrier trapping and recombination reveal that in colloidal particles (ca. 120 Å), the electron trapping time is less than 20 ps while that for hole trapping required on average 250 ns.[43] At high light fluxes, recombination followed second order kinetics, but became first order at low charge carrier occupancies, with intraparticle electron recombination requiring about 30 ns. Such charge accumulation makes possible the observation of multiple electron transformations, as in the sustained four quanta photodecomposition of water.[44]

2.3.2. Kinetics

Detailed kinetic profiles can be obtained by a number of routes besides conventional flash photolysis:[43–48] microwave conductivity,[49] laser flash-coulostatic profilometry,[50] measurement of laser-induced transient open circuit photovoltages,[51] and luminescence quenching[52] have all been used effectively to characterize the rates of interfacial electron transfer to (or from) a photogenerated electron-hole pair. Flash photolysis of optically transparent colloidal suspensions allows direct determination of quantum yield for both oxidative and reductive charge trapping by interfacial electron transfer[47] and rates of subsequent dark reactions that consume the photogenerated surface-bound radicals. We have found, for example, that the dependence of the quantum yield for electron trapping by adsorbed methyl viologen on colloidal CdS depends on the laser pulse width and incident intensity.[45] The experimental data were successfully simulated with a numerical kinetic analysis,[46] allowing for direct determination of the individual rate constants. On colloidal TiO_2, differences in quantum yield for 30-ps and 10-ns pulses indicated that both electron and hole trapping occurred within several tens of picoseconds, with trapped electron-hole recombination occurring on a much slower timescale. Since no such dependence was observed on colloidal CdS, the nature of the surface states and the energetics for recombination must sensitively influence the relative kinetics observed at interfaces.[45]

We have also recently shown that laser flash photolysis, coupled with diffuse reflectance spectroscopy, can be used with optically opaque powders to directly characterize transient kinetics on powders without having to extrapolate from the optically transparent, but less practical, colloidal suspensions.[53] A two-parameter model originally developed by Albery et al.[54] to describe kinetics in micelle systems, when applied to a diffuse reflectance absorptive transient, showed that the kinetics for the decay of $(SCN)_2^{\bullet -}$ on powdered TiO_2 followed first-order kinetics with a distribution of rate constants (Figure 5).[53] The average rate constant increased with increasing adsorbate concentration and decreases with increasing pH and inert electrolyte concentration, as would be consistent with a model involving a charged particle surface. The much faster decay

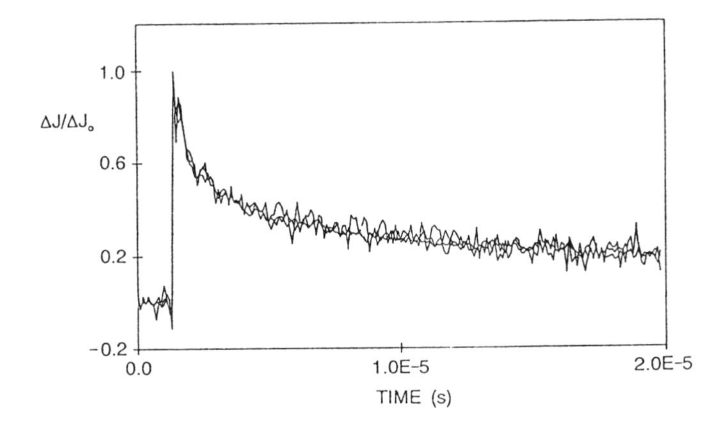

Figure 5. Decay of $(SCN)_2^-$ Signal Intensity Formed on Pulse-Irradiated TiO_2 from Diffuse Reflectance Measurements. Rough Curve is Experimental Data. Smooth Curve Represents First-Order Fit With a Distribution of Rate Constants According to the Albery Kinetic Model.[53]

(15 μs) observed on TiO_2 powder than on a freshly prepared colloidal suspension (300 μ) was attributed to the reduced dimensionality encountered as a surface-bound intermediate diffuses along the surface of a particle that is large on a molecular scale.[53] Attempts to use this technique to detect $OH^{\bullet}$ adducts in the titanium dioxide photocatalyzed reactions of iodide, potassium hexacyanoferrate, 2,4,5-trichlorophenol, methyl viologen, tetrachloroplatinum(II), tris-(1,10)phenanthroline iron, N,N,N′,N′-tetramethyl-*p*-phenylenediamine and thianthrene revealed only products of direct electron transfer.[55] Since the rates of reaction of the corresponding $OH^{\bullet}$ adducts are known from pulse radiolysis experiments in homogeneous solution,[56,57] this observation implies that electron transfer reactions, possibly involving photogenerated $OH^{\bullet}$, dominate the early stages of reaction, at least when compared with rates of hydroxyl radical addition to these substrates.

Steady-state luminescence measurements have similarly shown that direct interfacial electron transfer dominates the chemistry of adsorbates.[58] The addition of unsaturated hydrocarbons such as 1-butene or 1-butyne onto TiO_2 causes an increase in the observable photoluminescence of the catalyst, presumably as a radiative probe for electron-hole recombination. The intensity of the luminescence depends strongly on the ionization potential of the adsorbate, with parallel enhancement of photoluminescence and rates of photocatalytic hydrogenation in water. It is likely therefore that both processes depend critically on binding of the adsorbate to the photocatalyst's surface.

2.3.3. Cocatalysts

The evolution of gases through multielectron routes requires a bifunctional photocatalyst: as cocatalysts, platinum is often used as the reductive site for hydrogen evolution from water splitting and ruthenium oxide for the oxygen evolution site.[43] The exchange current density for hydrogen evolution peaks with transition metals of metal–hydrogen bond strengths of about 15 kJ/mol, e.g., Pt, Rh, Ir, Re,[59] with electrocatalytic activity being directly related to fractional *d*-character of the deposited metal. This requirement for a transition metal cocatalyst appears to derive from the need for an intermediate value for adsorption energy since both formation and cleavage of the metal–hydrogen bond must occur at reasonable rates if an acceptable catalytic turnover is to be realized. The efficiency of a given metal as a hydrogen evolution catalyst can be gauged by generation of a reducing relay from photosensitization in homogeneous solution in the presence of the metal hydrosol, prepared either chemically or radiolytically.[60] The effect of metal doping in enhancing the conversion efficiency of water cleavage has also been widely studied.[61,63]

These metallic cocatalysts can be conveniently deposited on the surface of a light-responsive semiconductor by photocatalyzed reduction of an adsorbed cation.[64] Even the position of deposition can be changed by chemical methods,[65] and the deposition of more than one metal can be accomplished.[66] We have shown, for example, that metals can be deposited exclusively inside a zeolite,[67] and, in particular, exclusively on a semiconductor constrained within a zeolitic support.[68] In the former case, this integrated metal-containing system could be photosensitized by an associated porphyrin to evolve hydrogen,[69] whereas in the latter, band gap irradiation of the included metallized semiconductor evolved hydrogen while oxidizing sulfide.[68]

While codeposition of a metallic catalyst provides a site to accelerate a desired reaction (and hence to improve the electrochemical fill factor), it can also have undesirable effects. A surface-deposited metal will introduce a new electrical contact, and therefore alter the barrier height in band bending, thus shifting the rest potential of the metallized semiconductor. It may also accelerate the rate of recombination of the photogenerated electron-hole pair or suppress the formation of surface-bound peroxotitanium sites. The presence of metal on the surface can also block light absorption, although reduction of the size of the metal particle to dimensions smaller than the wavelength of incident light permits the preparation of transparent films.[70] The expense of the most effective metal catalysts for large-scale operation is also non-negligible.

It would therefore be desirable to develop nontransition metal catalysts for hydrogen evolution and to explore new semiconductor photocatalysts on which the effects of the metal cocatalyst can be more effectively controlled. Our group, for example, found that ZnS can act as an effective catalyst for hydrogen evolution when deposited on a metal chalcogenide semiconductor.[71] The ZnS·CdS catalyst was as active for H_2 generation as Pt/CdS. Tungsten disulfide could also be used in this capacity.[72] Silica supported WS_2 could be prepared as hexagonal crystallites by pyrolyzing WO_3 adsorbed on silica at 300°C. The WS_2 supported on silica was more active than Pt/SiO_2 for catalytic hydrogen evolution both in the dark and under visible light illumination (with CdS or fluorescein as photosensitizers), although a graphite electrode immersed in a Pt/SiO_2 slurry could evolve hydrogen at a more positive applied potential than in a WS_2 slurry.[72] The longer induction period observed for gas evolution on WS_2 was associated with appreciable hydrogen adsorption by the catalyst and/or the reaction of atomic hydrogen with surface-accessible W^{4+} ions. The catalyst's dark catalytic activity could also be increased by doping with nickel, although the efficiency of dye sensitization was reduced by the presence of nickel.[73] The suggested explanation focused on the role of surface-bound nickel atoms in influencing the surface geometry and electronic properties of the catalyst.

3. ORGANIC REACTIONS ON IRRADIATED SEMICONDUCTORS

Having seen an overview of the principles operative in semiconductor-mediated photoelectrochemistry, the reader is now ready to consider how this reactivity can be exploited in controlling organic transformations. Various aspects of this topic have been treated elsewhere,[74–82] so we focus in this section on mechanistic features of these conversions. The coverage here is restricted to organic photoelectrochemical reactions since the more thoroughly studied reactions of simple inorganic compounds, e.g., water splitting, carbon dioxide reduction, etc., have been extensively reviewed elsewhere.[14,83–89]

As we noted earlier, the highly positive potentials of metal oxide semiconductors make them very active in inducing oxidative transformations of appropriate adsorbates. Most of the well-characterized reactions in organic photoelectrochemistry have exploited this high reactivity to activate oxygen[90] and induce photochemically driven oxygenations or oxidative cleavages in which the first step is electron transfer from the adsorbed organic donor to the surface-confined hole of a photogenerated electron-hole pair. Many classes of organic compounds can function as this donor, and this section will survey some of the transformations thought to occur by this direct electron transfer interaction.

In contrast, the most well-characterized photoelectrochemical reduction involves the production of hydrogen gas, or adsorbed hydrogen atoms. In the presence of an appropriate unsaturated organic molecule, these reducing equivalents can be taken up, effecting a net reduction. It is also possible to directly reduce organic molecules that are heavily substituted with strongly electron-withdrawing groups, thus diminishing their reduction potentials below the conduction band edge. This direct reduction is of significant importance in the photocatalyzed decomposition of polyhalogenated compounds, a problem of serious environmental concern. Examples of both photohydrogenations and reductive cleavages will be cited.

Both net oxidations and reductions are initiated by interfacial electron transfer, and if the resulting surface-bound radical ion rearranges and participates in back electron transfer before it desorbs from the catalyst surface, net rearrangement reactions can be accomplished, via reversible interfacial electron transfer on the illuminated semiconductor surface, without net oxidation or reduction chemistry having taken place. Because the operation of this mechanism makes special kinetic demands on the adsorbed redox participants, fewer reactions of this type have so far been investigated.

3.1. Solvent

Most organic photoelectrochemical studies seeking to explore direct surface-mediated electron transfer have been conducted in inert solvents, i.e., those that are oxidized at potentials beyond the band edges of the semiconductor chosen. The most commonly used solvents are acetonitrile and methylene chloride, although benzene and several others have been employed less frequently. Kinetic differences observed in the rates of geometric isomerization of substituted stilbenes on irradiated suspensions of CdS, TiO_2, and ZnO in aprotic solvents have been ascribed to perturbation of the substrate adsorption equilibrium, and hence to the alteration of surface properties accompanying this physisorption.[34b]

When such reactions are conducted in protic solvents, e.g., alcohols or water, the solvent itself competes with the desired substrate for surface-active sites. This competition not only reduces the efficiency of the reaction of interest, but also introduces mechanistic complexity by allowing routes for generation of other reactive radicals. For example, photocatalyzed oxidations conducted in water are often dominated by the photochemical formation of the hydroxyl radical. This species is highly reactive and therefore unselective. Under such conditions, trapping of intermediates may be difficult,[55] with the main product being complete mineralization. In the reaction of benzene, for example, as a saturated aqueous suspension of TiO_2, only traces of phenol can be detected, with carbon dioxide being formed as the major product of this indiscriminate photooxidation.[91,92] Hydroxyl radicals formed by irradiation of suspended TiO_2 can be characterized by photoluminescence quenching and esr spectroscopy,[93] and their ubiquity in photoelectrochemical reactions conducted in water is now well-established.[94] This reactivity can be specifically exploited in devising new routes to initiate photo-Fenton chemistry,[95] but it makes selectivity difficult to attain in aqueous solution.

The effect of solvent can be probed by the use of solvent additives that can compete on a more equal concentration basis with the substrates of interest. The addition of small amounts of halogenated alcohols to an acetonitrile suspension of TiO_2 containing α-methyl styrene caused an increase in the rate of reaction and shifted the observed product distribution away from simple oxidative cleavage.[96] The kinetic effect is caused by suppression of electron-hole recombination. It can be explained as enhanced electron trapping deriving from higher local concentrations of oxygen near a surface rich with trihalomethyl groups. The corresponding nonhalogenated alcohols inhibited the reaction, presumably by establishing a competition between the additive and the substrate for adsorption sites without a parallel increase in electron-trapping efficiency. Consistent with this argument, other highly fluorinated nonalcoholic additives cause a more modest rate of increase, as would be expected from the lower equilibrium association of these compounds. Thus, conditions for optimal photooxidative chemistry can be obtained by controlling the solvent environment at the critical interface.

3.2. Photooxidation/Photodehydrogenation

Because the photocatalytic activation of organic molecules on metal oxide and metal chalcogenide surfaces is so relevant to thermal-catalyzed redox reactions on the same surfaces, it is important to make a distinction between direct mediated single electron exchanges (oxidation/reduction) and those that involve hydrogen atom transfers (dehydrogenation/hydrogenation). Mechanistically this is a crucial distinction, for the former route will involve charged intermediates, i.e., radical ions, and will be responsive to the redox properties of the adsorbate. The latter route will involve neutral radicals, at least initially.

Since the chemical properties and adsorption of these two types of reactive intermediates will be quite different, an important mode of reaction control involves the ability to change conditions to specify one reaction route over the other. In practice, this may be difficult. For example, the deposition of a metal island on the surface of a photoresponsive semiconductor will enhance the oxidative route by improving electron-hole separation, but it will also improve the efficiency of the dehydrogenation since the metal island can also act as a catalytically active site for hydrogenation–dehydrogenation equilibria. Oxygen is often employed to differentiate these pathways. Since dehydrogenation typically involves metal-bound hydrogen atoms, which can be scavenged by oxygen to produce water, the presence of oxygen effectively inhibits hydrogen atom transfer. Thus, reactions that occur in the absence of oxygen or other electron scavengers are probably dehydrogenations, although the converse is not necessarily true.

Oxygen will also enhance the facility of electron transfer, however, since it can act as an effective electron trap to promote separation of the photogenerated electron-hole pair and thus inhibit charge recombination. Its low reduction potential makes it quite attractive as this electron trap, but other reagents can also be similarly used. Direct reduction of methyl viologen (an organic dication) can be observed spectroscopically,[47] and the use of carbon tetrabromide[97] and Ag^{+}[98,99] in this capacity has been inferred from product analyses.

3.3. Organic Photoelectrochemical Oxidations

Perhaps the most striking feature of photoelectrochemical oxidations occurring on irradiated semiconductor surfaces is their ubiquity. Nearly any organic substance can be oxidized under these conditions, either directly or though the mediation of a solvent-derived radical. One memorable paper describes the use of polyvinyl chloride, algae, protein, dead insects, and animal excrement as sacrificial electron donors for photoinduced electron transfer.[100] Semiconductor-mediated photooxidations have even been proposed recently as a principal route for prebiotic functionalization of simple molecules and as ecologically significant parts of the nitrogen cycle.[101] Although organic photoelectrochemistry is far less thoroughly explored than transformations involving water splitting, the literature is still too voluminous for an exhaustive review. The following examples typify, however, the reactions that have been described so far. The considerations discussed above would suggest that substrates bearing easily oxidizable substrates would react by interfacial electron transfer with the substrate of interest functioning as an adsorbed donor (Eq. 2).

3.3.1. Mechanism

Any mechanistic study of a photocatalyzed electron transfer reaction will begin by establishing the role of light. This is usually accomplished by establishing the correspondence of the photoaction spectrum[82] with the absorption spectrum of the photocatalyst. The efficiency of a photoelectrochemical reaction is often expressed as product formation per incident photon rather than as a true quantum yield because of the difficulty in measuring light absorption in nonhomogeneous, optically opaque media. Structure–activity profiles, product analyses, spectroscopic characterization of intermediates, temperature effects, etc. are evaluated in the same way in these heterogeneous suspensions as in standard studies in homogeneous solution.

We will consider the mechanism of two reactions in detail: the oxidation of alcohols and the oxidative cleavage of olefins on irradiated semiconductor surfaces. These two examples are chosen since they involve the extreme types of initial coordination: either through hydrogen bonding or the lone pair of the OH group in the alcohol or through π coordination in the olefin. In this comparison, we can see how proton transfer affects the efficiency of organic photoelectrochemical conversions.

3.3.1.1. Photoinduced Oxidation of Alcohols. Photochemical excitation of a metal oxide semiconductor suspended in neat alcohol or in an inert solvent containing an alcoholic substrate effects conversion to the corresponding carbonyl compound (Eq. 12).

$$\mathrm{RR'CHOH} \rightarrow \mathrm{RR'C{=}O} \tag{12}$$

Most such studies have employed alcohols as sacrificial reagents to test catalytic efficiency in the production of hydrogen gas,[102–106] and less emphasis was placed on product analysis or methods to achieve chemical selectivity. In some cases, complete mineralization to CO_2 and water was the primary goal.[103] Alcohol structure did affect the quantum yield, with water being oxidized about four times less efficiently than ethanol, and methanol and ethanol being more efficiently oxidized than other primary alcohols.[105]

Our earlier conclusion that the higher relative reactivity of primary over secondary or tertiary alcohols[21,22] was attributable to differences in adsorption seems to be borne out in more complex systems as well. Photocurrents detected in the photocatalyzed oxidation of polyols were found to depend on molecular structure, i.e., chain length and number of hydroxyl groups,[107] and the magnitude of the current appeared to relate to the concentration of OH groups on the surface (and hence the solution pH) to which the polyol was adsorbed.[108] If D_2O and ethanol were allowed to compete for the same catalytically active site, D_2 was found to be the major product (88%).[109] In this study the ratio of hydrogen to methane evolved was about 14, indicating that direct mineralization to CO_2 and water proceeded about four times faster than routes involving photo-Kolbe decarboxylation of acetic acid.

The overall efficiency of the reaction was significantly affected by the reaction conditions. The degree of platinum loading exerted a strong influence on the efficiency of hydrogen evolution, with optimal results obtained with a loading on TiO_2 of 0.1 to 1 wt%, in which the platinum islands had diameters of approximately 2 nm.[104–106,110] Besides Pt, other noble metals and transition metal catalysts[111] and surface bound metal oxides, e.g., vanadium or molybdenum oxides,[112] could be used to catalyze the reductive half-reaction by acting as electron traps for the photogenerated electron-hole pair, although the metal oxides decreased photochemical activity. Alkalai metal salts as additives appeared to enhance the relative activity of water in aqueous methanol,[113] although the use of metallic nickel as a support caused the inhibition of the oxidation of 2-propanol in the absence of water.[114,115] Fiberoptic monitoring established that photoactivity occurred at acidic Ti^{4+} sites,[116] with surface binding occurring at a non-associated hydroxyl group.[117]

With secondary and tertiary alcohols, the photoinduced oxidation was more difficult, with carbon–carbon bond cleavage products and products derived from radical coupling being isolated along with simple oxidation products.[118] The radical products are thought to derive from $OH^{\bullet}$ secondary reaction.[119] Dehydration was also found to occur in secondary and tertiary alcohols, with skeletal cleavages occurring on photoinduced activation of 3-methylbutanol.[120] Selectivity in the alcohol oxidations could be characterized by kinetic methods:[121] a rate expression consistent with Langmuir–Hinshelwood adsorption was observed for simple primary alcohols,[122] but a more complex kinetic model was necessary at higher concentrations.[123] The 8 : 1 preference for formation of butanal over butene and the independence of butene yield on adsorbate composition in the photooxidation of 1-butanol were explained as occurring at least two active sites: one at unsaturated surface O^{2}-sites displaying mixed first-order Langmuir–Hinshelwood kinetics and a second at another site following Eley–Rideal kinetics.[124] The latter was inferred from the observed continuous rate increase with initial alcohol concentration, but with a smaller slope at higher concentrations. A Mars–van Krevelen rate expression (Eq. 13) provided excellent fit for the observed data:

$$r = 1/k_{red[ROH]} + 1/k_{ox[O_2]} \tag{13}$$

This assignment of separate sites for different photoactivated modes was also observed in the photoinduced oxidation of 2-propanol on acidic TiO_2 suspensions in the presence of silver ion as electron acceptor.[125]

Thus, the mechanism of alcohol oxidation requires first a pH sensitive adsorption onto the photocatalyst surface, with the extent of deprotonation of both the surface and substrate varying with reaction conditions. Photogeneration of an electron hole-pair allows for either direct single electron transfer from alcohol bound at the active site or surface activation to an intermediate surface-bound hydroxyl group (Scheme 1). In the first route the radical cation rapidly deprotonates, forming an adsorbed radical, oxidation and deprotonation of which lead to oxidation product.

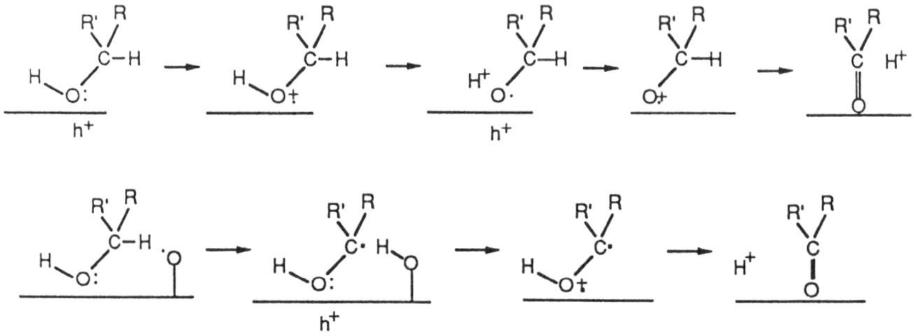

Scheme 1. Photocatalytic Dehydrogenation of Alcohol.

Desorption from the surface then completes the observed reaction. The mediated reaction takes place through hydrogen atom abstraction from the alcoholic substrate bound adjacent to the site of the activated surface-bound reagent. This may be a hydroxyl radical,[119] a surface oxide, or a titanium peroxy radical,[126] the latter deriving possibly from attachment of superoxide formed via electron trapping by adsorbed oxygen.[127] The resulting radical can then be further oxidized and desorb. The positions of the deprotonation from the first route and the hydrogen abstraction in the second may not be the same, since the deprotonation is governed by relative acidity and the hydrogen abstraction will be determined by relative bond strengths and access to the abstracting reagent. Thus, product selectivity (and observed kinetics) will depend on the surface characteristics of the photocatalyst, especially on surface pH.

A specific example can be seen in the proposed mechanism for the TiO_2 photocatalyzed oxidation of 2-methoxyethanol (Eq. 14).[128]

$$HOCH_2CH_2OMe \xrightarrow{TiO_2^*} HCO_2Et + EtOH + CH_3CHO + CO_2 \quad (14)$$

Here the major product, ethyl formate, is thought to arise from hydrogen abstraction by surface bound superoxide (Scheme 2), initiating formation of a surface formate.

Analogous processes might be expected in the photoxidation of phenol. Surface adsorption has been established on TiO_2 by direct imaging by scanning tunneling microscopy.[129] When the photooxidation was conducted on aqueous acidic TiO_2 suspensions, the hydroxyl radical was implicated as the reactive species since oxygen was incorporated.[130] Continued photolysis lead to ring-cleaved aldehydes, acids, and CO_2. On

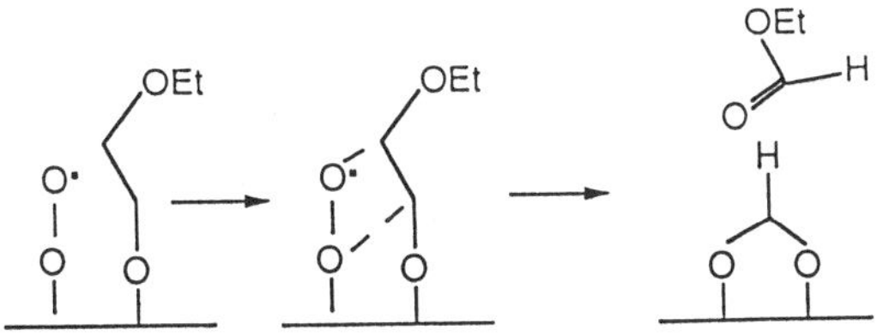

Scheme 2. Surface Formate Formation.[128]

ZnO, the yield of oxidation product was found to decrease with increasing phenol concentration and pH and with decreasing oxygen concentration, leading the authors to propose the involvement of $OH^{\bullet}$ reacting with homogeneously dispersed phenol in dilute solutions and direct capture by photogenerated holes by phenol in more concentrated solutions.[131] The hydroxyl radical is also thought to be involved in the photocatalyzed hydroxylation of chlorophenols.[132] As with the simple alcohols, Langmuir–Hinshelwood kinetics were observed, implying that the reactant and solvent compete for the same active site or that both reactant and solvent are adsorbed at a nonactive site.[133]

3.3.1.2. Photocatalyzed Oxidative Cleavage of Olefins. The oxidative cleavage of olefins is one of the most thoroughly studied photoelectrochemical conversions. We illustrate the reaction with 1,1-diphenylethylene (DPE), a substrate whose dark oxidative chemistry is known. Irradiation of a suspension of TiO_2 in anhydrous aerated acetonitrile containing DPE produces benzophenone in essentially quantitative yield (Eq. 15).[134,135]

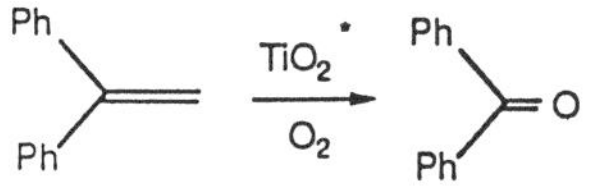

$$Ph_2C{=}CH_2 \xrightarrow[O_2]{TiO_2^{\bullet}} Ph_2C{=}O \quad (15)$$

At shorter irradiation times, a small amount of epoxide and α-carbonyl compound (Eq. 16) can also be isolated. Both intermediates are ultimately converted to benzophenone with further irradiation.

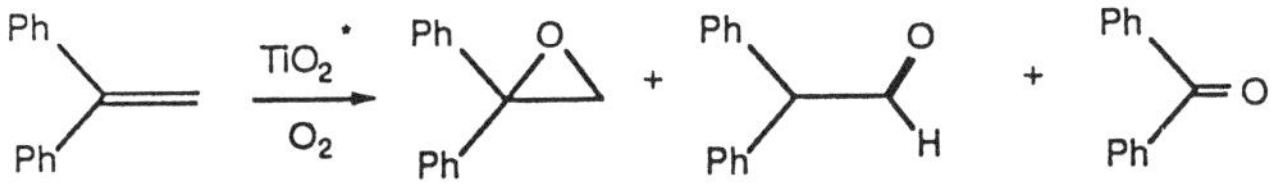

$$Ph_2C{=}CH_2 \xrightarrow[O_2]{TiO_2^{\bullet}} Ph_2C(O)CH_2 \text{ (epoxide)} + Ph_2CH{-}CHO + Ph_2C{=}O \quad (16)$$

In this system, the relevant energy levels are shown in Figure 6: the valence band of TiO_2 lies about 0.6 V positive of the electrochemical peak potential for the oxidation of DPE, so that hole trapping by adsorbed DPE can exothermically generate the adsorbed cation radical. The conduction band of TiO_2 is nearly isoenergetic with the reduction potential of oxygen,[135a] so electron trapping by adsorbed oxygen will form surface-bound superoxide. Collapse of these radical ions through a four-membered transition state could potentially form a dioxetane. Although no evidence for intermediate formation of this species could be obtained, a control experiment demonstrated that this compound, independently synthesized, when stirred in the dark while suspended on TiO_2 is quantitatively converted to benzophenone and formaldehyde.[134] This dark reaction exploits the Lewis acidity of surface titanium ions. Attempts to observe chemiluminescence during the dark reaction or during the photocatalytic oxidation were unsuccessful. The dioxetane is thus a permissive intermediate in this conversion.

Back electron transfer within the photogenerated adsorbed radical ion pair is sufficiently exothermic (> 2 eV, i.e., >50 kcal/mol) to form singlet oxygen. Since singlet oxygen does lead to benzophenone with DPE, its involvement in this sequence cannot be discounted summarily. We must therefore consider the electron transfer initiated sequences shown in Scheme 3 as routes to the observed cleavage products.[82] Respectively, these routes consider reaction of the neutral olefin with superoxide or singlet oxygen and the trapping of the olefin cation radical by ground state (triplet) oxygen or superoxide. The last pathway involves sequential trapping of the cation radical with starting material and with triplet oxygen.

The importance of the top route can be tested by treating the starting material in anhydrous acetonitrile with potassium superoxide solubilized by 18-crown-6 ether. After prolonged stirring under these conditions, the starting olefin could be recovered unchanged.[134] Nor did superoxide added to the solution in contact with the photocatalyst change either the rate of oxygenation or the product distribution observed at short irradiation times. Phenylglyoxylic acid, a well-characterized trap for superoxide,[136] also produced no observable effect on the reaction. Thus, superoxide, if involved in the reaction, must react on the surface of the photocatalyst rather than by diffusing from the surface to interact in solvated form in the bulk.

The possible involvement of singlet oxygen in such oxidative cleavages is made unlikely by the contrasting reactivity observed in the semiconductor-mediated oxygenations and those initiated by singlet oxygen. It is known, for example, that *trans*-stilbene fails to react with

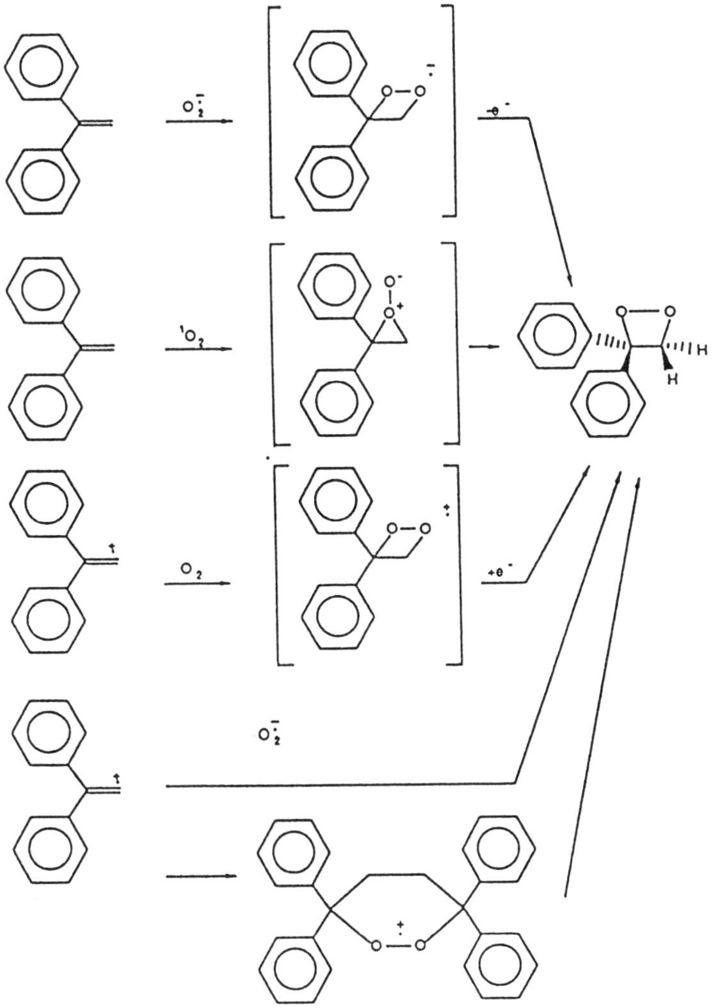

Scheme 3. Proposed Routes for Olefin Oxygenation.[82]

singlet oxygen except through electron transfer routes.[137] Similarly, singlet oxygenation of tetramethylethylene proceeds via an ene reaction to form allylic hydroperoxide,[138] whereas photoinduced electron transfer of this same substrate leads to acetone, presumably in as route parallel to that observed with DPE.[134] Furthermore, despite several attempts to do so, no unambiguous evidence for the formation of singlet oxygen on irradiated metal oxide surfaces has been reported.

The remaining routes in Scheme 3 all involve intermediate cation radical formation. *Para*-substitution of the phenyl groups in DPE with electron-donating and -withdrawing groups allows for a test of electronic contributions to the transition state leading to the intermediate formed in the rate-determining step of the reaction. A plot of the relative rate of photooxygenation (under identical reaction condition) for a series of such substituted DPEs against σ^+ was linear with a ρ value of -0.56.[139] This slope is identical to that observed in a plot of the oxidation potentials of this series of hydrocarbons with σ^+, and is similar to ρ values reported for other reaction proceeding through intermediate cation radicals.[140]

Although this negative slope indicates that the rate-determining step of this conversion involves the development of positive charge, its value does not permit unequivocal assignment of this step to interfacial electron transfer or to a subsequent step that localizes positive charge at the 1-position. This localization might occur as the photogenerated cation radical is trapped by oxygen or superoxide at the β-position, for example. MO calculations indicate that the change in charge density developed at C-1 is similar in the formation of the cation radical (which is quite delocalized) and in bond formation at C-2 in the cation radical.

Direct evidence for transient formation of the olefin cation radical is also available from flash photolysis experiments. For example, the transient formed by flash photolysis of a colloidal TiO_2 suspension in the presence of *trans*-stilbene is identical both in absorption maxima and in lifetime to the same transient produced pulsed radiolytically in the same medium.[47] If it is possible to generalize behavior from colloidal semiconductor suspensions to the corresponding powders, these observations provide direct evidence that photoexcitation does indeed produce adsorbed cation radicals.

Although none of these mechanistic arguments alone is compelling evidence for the involvement of a photogenerated cation radical, the composite picture surely is consistent with this conclusion, particularly in view of its correspondence with reasonable predictions from thermodynamic considerations applied to semiconductor band theory. Thus, the three remaining routes of Scheme 3 are all quite reasonable.

The third route in Scheme 3 involves trapping of the cation radical by ground state oxygen, a route that has been observed upon electrochemical generation of highly hindered olefins to form dioxetane, epoxide, and α-carbonyl products.[141] That these are the same products isolated from DPE at short irradiation times may not simply be fortuitous. If this route is operative, the question of whether one or two C—O bonds are formed in the rate-determining transition state remains. MNDO calculations show that single bond formation is much more favorable than the concerted two bond route.[142] The zwitterion formed can then collapse either to a dioxetane or peroxirane intermediate from which epoxides can be obtained. Thus, if the surface bound cation radical is captured by ground state oxygen, the route shown in Scheme 4 should be followed.

If the photogenerated cation radical could diffuse along the surface, however, it could also encounter the superoxide anion radical formed by interfacial electron trapping by adsorbed oxygen. Such diffusion of adsorbed reactive intermediates is a common feature in the photoreactivity of adsorbed molecules,[143,144] and the facility of a given reagent in finding a coreactant in this two-dimensional migration will depend on the fractal character of the surface, which can be used to predict a power law dependence of the photophysical observable (diffusion rate, lifetime, etc.).[145]

In principle, the rates of cation radical trapping by oxygen and superoxide could be distinguished kinetically, for the former reaction occurs at a rate about three orders of magnitude slower than the latter (which is diffusion controlled) in homogeneous solution.[146] In practice, however, this requires distinguishing rates of reaction from much slower rates of surface diffusion that probably govern such events. Furthermore, as kinetic studies employing diffuse reflectance spectroscopy for kinetic analysis have shown,[53] kinetics on the surface of an irradiated photocatalyst are complex, presumably because of differential distances between photocatalytically active sites in this heterogeneous medium. Thus, the fourth route of Scheme 3 also remains viable.

Each of these routes is projected to proceed through a dioxetane, decomposition of which would give formaldehyde along with the observed benzophenone. The high volatility of formaldehyde, together with its high photoactivity under photooxidative conditions, makes its quantitative detection difficult. This in turn makes the mechanistic choice

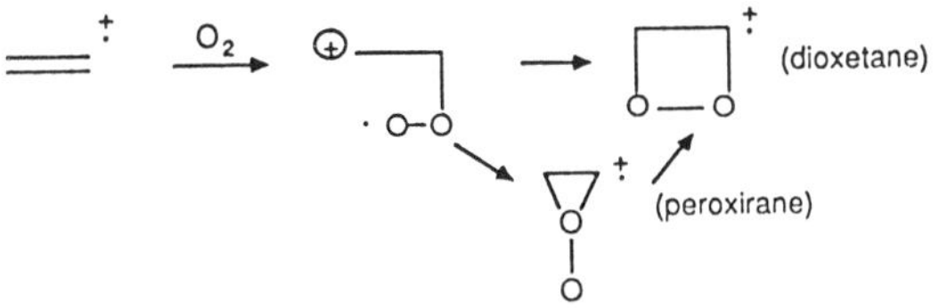

Scheme 4. Oxygenation of the Ethylene Cation Radical.[142]

between route (5) and routes (3) and (4) in Scheme 3 ambiguous. Ethylene, the expected side product of route (5), is also volatile and photoelectrochemically active. In Eq. 15, nonetheless, traces of formaldehyde can be found in the product mixture, but no evidence for formation of ethylene was evident. Furthermore, in the semiconductor-mediated oxidative cleavage of unsymmetrically substituted *trans*-stilbenes, nearly equal quantities of benzaldehyde and the substituted benzaldehyde were obtained, with no evidence of metathesis, as might have been expected had route (5) been followed.

Identifying the critical reactive intermediate in such reactions is helpful, but often insufficient, in determining the ultimately observable products. With diphenylethylene, for example, completely different products are obtained from the cation radical with photoelectrochemical activation than by other means. Shown in Scheme 5 is a summary of different reactivity from the cation radical formed photoelectrochemically, by electrochemical oxidation on an inert metal electrode, and by a homogeneously dispersed single electron oxidant.[78] The dimerization occurring on a poised metal electrode derives from the high concentration of radical cations formed in that environment. The photoelectrochemical surface, in contrast, forms a much lower steady-state concentration of the reactive intermediate, and in the presence of an active coreactant (O_2 or superoxide) in the immediate environment, undergoes oxidative trapping before desorbing into the contacting solution. The homogeneous reaction is more complex: the major product can be obtained in only about 25% yield. Nonetheless, its contrasting reac-

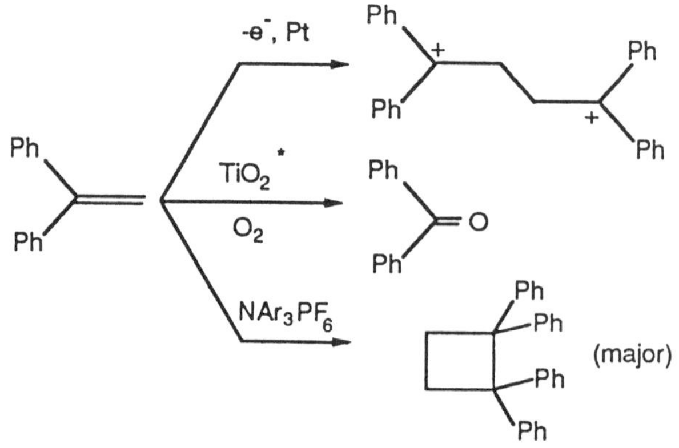

Scheme 5. Divergent Pathways from a Common Cation Radical.[75]

tivity from that observed with the same intermediate formed on a photoactive surface is striking.

Although this mechanism (photogeneration of surface-adsorbed cation radicals) is also reasonable with other oxidizable alkenes, these same considerations would predict that olefins whose oxidation potentials lie positive of the semiconductor valence band edge (about +2.5 V) would take place by a different course. This expectation is borne out in the reactions of several nonarylated olefins. With β-pinene, for example, both allylic oxidation and oxidative cleavage are observed. Presumably this route does not involve surface-adsorbed cation radicals but rather the intermediacy of hydroxyl radicals and routes similar to those discussed earlier in the photooxidation of alcohols. Similarly, the photooxidation of propylene on irradiated metal oxide suspensions has been reported to give CO_2 as well as products of intermediate oxidation level.[147] Selectivity in the photooxygenation is dependent on the catalyst and the reaction conditions, with acetaldehyde formation (78%) dominating over propylene oxide formation (18%) and allylic oxidation (to allylic alcohol: 3%) on suspended TiO_2.[148]

Further evidence that these reactions proceed through redox- activated surface-oxide groups derives from the isomerization chemistry observed with simple alkenes. Hydrocarbons on irradiated semiconductor suspensions in aerated inert solvent suffer bond cleavages, either of C–H, C–C, C=C, or C≡C bonds with concomitant incorporation of oxygen. With 2-butene, for example, both geometric isomerization and double bond shifts are attained with photocatalytic activation of semiconductor suspensions.[149] The isomerization of 2-butenes to 1-butene is contrathermodynamic and is obviously driven by the photoactivation. The presence of oxygen or NO inhibits the reaction, and reactivity is assigned to a Ti^{3+}–O^- charge transfer state with surface O^- sites being responsible for the double bond shift. Separate reaction routes have been proposed for each of four modes of reactivity: geometric isomerization at these Ti^{3+}–O^- sites, double bond isomerization by surface bound OH^- or $OH^\bullet$, hydrogenolysis by surface bound radicals ($OH^\bullet$ or $H^\bullet$), and hydrogenation by $2H^\bullet$.[150] Similarly, with alkanes photocatalyzed oxidation reactions seem to derive from intermediate radicals rather than through cation radicals formed via surface- mediated interfacial electron transfer, since oxygenation at stable radical sites dominates over oxidative cleavage.[151–153] The careful choice of the photocatalyst makes possible some degree of control of the divergence between simple oxygenation and complete mineralization.[153]

The onset of differential reactivity between the surface-oxide-mediated reactions and those involving cation radical formation is governed by simple electronic structural factors within the molecule. Further unsaturation (beyond a simple double bond) seems to ease oxidation sufficiently to allow for efficient hole trapping. Thus, dienes[154] and alkynes[155] are oxidatively cleaved in a pathway closely resembling that discussed above for 1,1-diphenylethylene. The isolated products seem to mimic those obtained by thermal routes for trapping of diene cation radicals by molecular oxygen.[156] Cyclization reactions characteristic of diene cation radicals can also be observed.[99]

In summary, we expect two major oxidative methods to be operative in the semiconductor-mediated photooxidation reactions of organic substrates. These involve either surface-activated oxide centers or direct interfacial electron transfer to generate adsorbed cation radicals. In either case, adsorption of the substrate to the semiconductor surface before photochemical excitation is necessary if reasonable efficiencies are to be observed. This, in turn, causes the initial oxidized intermediates to be formed as surface-associated reagents, so that their subsequent reactions can be kinetically dictated by the local environment at the surface.

3.3.2. A Survey of Functional Group Oxidative Reactivity

These mechanistic features are also characteristic of other functional group transformations on irradiated semiconductor suspensions. Representative examples of some of these, with minimal descriptive comment, are offered to the reader in this section.

Arenes are photooxidized either by side chain oxidation as in the formation of benzyl alcohol (and benzaldehyde) from toluene on irradiated anatase suspended in the neat liquid (Eq. 17).[151]

$$C_6H_5CH_3 \xrightarrow[PhCH_3,\ O_2]{TiO_2^*} C_6H_5CHO \qquad (17)$$

In the gas phase, aldehydes are produced,[157] while both oxidation and coupling products are observed when the photocatalyzed reaction is conducted in water.[158] Benzene-saturated aqueous suspensions of TiO_2

produced indiscriminate oxidation.[91,92] With large ring systems, e.g., substituted naphthalenes, oxidative cleavage of the ring is observed (Eq. 18).[159]

$$\text{1-methylnaphthalene} \xrightarrow[\text{CH}_3\text{CN},\ \text{O}_2]{\text{TiO}_2^{\bullet}} \text{2-acetylbenzoic acid } (\text{C}_6\text{H}_4(\text{COCH}_3)\text{CO}_2\text{H}) \tag{18}$$

Regioselectivity in the cleavage is consistent with the involvement of a cation radical, with that ring bearing the greater charge and spin density being most effectively oxygenated.

In considering oxygen-containing substrates, we have seen that alcohols are converted to the corresponding carbonyl compound, the efficiency of which is related to structural factors controlling adsorption.[160] These carbonyl compounds, in turn, can be converted further to the corresponding carboxylic acid.

We have also seen that photo-Kolbe reactivity occurs on irradiated semiconductor suspensions, effecting decarboxylation and, hence, a fragmentation of the carbon skeleton of the starting material. This reaction (Eq. 19) is historically one of the first involving photoelectro-

$$\text{Me}_3\text{C}\text{–CO}_2\text{H} \xrightarrow[\text{H}_2\text{O},\ \text{O}_2]{\text{TiO}_2^{\bullet}/\text{Pt}} \text{Me}_3\text{C}\text{–H} \tag{19}$$

chemical catalysis to be investigated mechanistically.[25,26,38,161] It differs from the electrochemical Kolbe in leading preferentially to simply reduced rather than coupled hydrocarbons, a manifestation of the chemical control available on the surface by virtue of back electron transfer to the adsorbed radical. Exhaustive oxidation of these and more complex carboxylic acids such as oxalate[162] and EDTA[163] can be used as a method for complete mineralization of pollutants.

Ethers can be converted to the corresponding ester (Eq. 20).[164]

$$\text{PhCH}_2\text{OCH}_3 \xrightarrow[\text{CH}_3\text{CN},\ \text{O}_2]{\text{TiO}_2^{\bullet}} \text{PhC(=O)OCH}_3 \tag{20}$$

The high selectivity for deprotonation α to the ethereal oxygen implicated initial formation of an adsorbed cation radical by trapping of the photogenerated hole. This α-deprotonation also leads to carbon–carbon coupling on irradiated ZnS suspensions (Eq. 21).[165]

$$\xrightarrow[H_2O,\ O_2]{ZnS^*} \quad \text{and diastereomers} \qquad (21)$$

Nitrogen-containing compounds are similarly oxidized. Primary amines, for example, can be converted to the corresponding imine, which, depending on concentration and solvent, can either be hydrolyzed or trapped by reaction with starting material. 4-Phenylbutylamine shows concentration dependence in which cleavage products are observed at low concentration and products characteristic of nucleophilic trapping, followed by photooxidative cleavage, at higher concentrations (Eq. 22).[166]

$$\xrightarrow[0.002\ M]{TiO_2^*, O_2} \qquad \xrightarrow[TiO_2^*, O_2]{0.1\ M} \quad NH_2 \quad H \quad CHO \qquad (22)$$

The intramolecular analog of this latter trapping can be seen in the reaction of α,ω-diamines (Eq. 23).[167] The alkylation of ammonia in the

$$NH_2 \quad NH_2 \xrightarrow{TiO_2^*/Pt} \quad NH \qquad (23)$$

presence of alcohol is likely to involve similar mechanisms.[168] Radical cations are presumably involved in the formation and reactivity of Schiff bases formed in parallel routes.[169]

The conversion of amides to imides by oxidation α to nitrogen (Eq. 24)[170] has also been reported in aqueous suspensions of TiO_2, presumably through hydrogen abstraction.

$$\text{N-acetyl-3-pyrroline} \xrightarrow[\text{H}_2\text{O},\ \text{O}_2]{\text{TiO}_2^{*}} \text{N-acetyl-3-pyrrolin-2-one} \qquad (24)$$

N–N coupling has been reported from the intermediates generated in the photooxidative N–H cleavage in aniline. Toluidine gives azo products,[171] while hydrazines are formed from ammonia.[18,172]

Photooxygenation of organic chalcogenides occurs readily. While diaryl sulfides undergo clean oxidation to the corresponding sulfoxide, and hence to the sulfone,[173] dibenzyl sulfides suffer carbon–sulfur cleavage, ultimately giving organic aldehydes (Eq. 25).

$$\text{PhCH}_2\text{SCH}_2\text{Ph} \xrightarrow[\text{CH}_3\text{CN},\ \text{O}_2]{\text{TiO}_2^{*}} \text{PhCHO} \qquad (25)$$

Sulfides further functionalized with halogens can also be cleanly converted to the corresponding sulfoxides on irradiated TiO_2 (Eq. 26).[173]

$$\text{CH}_3\text{SCH}_2\text{CH}_2\text{CH}_2\text{Cl} \xrightarrow[\text{O}_2]{\text{TiO}_2^{*}} \text{CH}_3\text{S(=O)CH}_2\text{CH}_2\text{CH}_2\text{Cl} \qquad (26)$$

Conversion of thioethers gives a complex mixture of cleavage products in which the Stevens rearrangement represents one significant pathway, while disulfide formation is observed in the photooxidation of thiols.[174] Phosphines can be photoelectrochemically oxygenated,[175] and phosphates, phosphites, and phosphidithioates can be decomposed to the elemental oxides on irradiated TiO_2 suspensions.[176]

Of great environmental importance is the decomposition of halogenated compounds, species which are widely dispersed and highly toxic. The complete mineralization of carbon tetrachloride on irradiated TiO_2 suspended in water represents a typical conversion,[177] but numerous other hazardous materials such as pentachlorophenol[178] or chlorinated dioxins[179] can also be photoelectrochemically degraded. The kinetics of such processes can be conveniently followed,[121] and their utility as routes for water purification is obvious.[180] Because of their practical importance, these applications have been extensively reviewed.[181]

Finally, the oxidation of simple inorganic ions and molecules can be used with organic substrates as a vehicle for establishing mechanism for these species. We have shown, for example, that cyclohexene can be used as a probe to distinguish formation of bromine atom, molecular bromine, and hydrobromous acid on the irradiated metal oxide surface.[182] Similar considerations have been applied to the trapping of oxidation products of Tl^{3+},[183] sulfur oxides,[184] and azide.[185]

3.4. Organic Photoelectrochemical Reductions

Reductions of organic substrates are much less common than are oxidations because of the only modest potentials of the conduction band edges of most common stable semiconductors. A few are notable, however. Most involve hydrogenations from hydrogen atoms formed on metallized semiconductor powders (by interfacial reduction of protons) or direct electron transfers to cationic, and therefore highly reducible, reagents such as the methyl viologen dication.[14]

In the photoelectrochemical hydrogenation of acetylene, a relay (a surface-bound molybdenum oxide unit) proved to be necessary for efficient reduction.[186] Coating the surface of an active semiconductor with a cationic surfactant increased the efficiency of oxidative cleavage, presumably by trapping the electron sufficiently to inhibit charge recombination.[187] Although a brief report has appeared that employs the photoelectrochemical reduction of $NADP^+$ in this capacity,[188] most reductions generate transferrable hydrogen atoms on metal islands deposited on the surface of the photocatalyst. For example, in the absence of oxygen, an acidic aqueous suspension of 3-cyclohexene-1,2-diacid could be hydrogenated at the double bond in preference to the decarboxylation observed in the presence of oxygen (Eq. 27).[78]

CO$_2$H, CO$_2$H $\xrightarrow[N_2,\ H_2O]{TiO_2^{\bullet}}$ CO$_2$H, CO$_2$H (27)

Band gap irradiation of CdS partially covered with platinum or rhodium led to acetylene hydrogenation with an efficiency about four times greater than for hydrogen evolution from the same mixture.[189] A sacrificial donor was needed for this reaction, which proved to be sensitive to pH and the method of deposition of the metal catalyst. As with standard catalytic hydrogenations, these photoelectrochemical hydrogenations occur in high yield with retention of the original carbon skeleton and with stereocontrol.[190]

Heteroatom-containing substrates can be similarly reduced. The photoelectrochemical reductions of thionine[191] and the N=N double bond of methyl orange,[192] for example, react by this route. Direct reduction of carbonyl double bonds has also been reported, leading to a photo-Cannizzaro reaction (aldehyde disproportionation) on irradiated ZnS sols.[193] The authors suggest that this reactivity may derive from a substantial negative shift of the conduction band edge on ZnS under these reported reaction conditions.

3.5. Photoelectrochemical Transformations with Reversible Electron Transfer as the Activation Step

The potential reversibility of electron transfer on an irradiated semiconductor surface makes possible the observation of isomerization reactions and pericyclic reactions, which are significantly kinetically accelerated as ion radicals. Most of such examples have involved the opening of strained rings as radical cationic retrocyclizations: some specific examples include retro [2+2] cycloadditions[34b, 194–197] and a retro [4+4] cycloaddition.[198] The photoelectrochemical dimerizations of phenyl vinyl ether[198] and *N*-vinylcarbazole[135b] on CdS appear to be similarly activated by interfacial electron transfer. In these reactions, a mechanism involving formation of the dimer both in the adsorbed state and in solution was proposed, with the adsorbed reaction dominating.[199] Some interfacial electron transfer could also be observed in the dark because of the presence of acceptor-like surface states on the surface of the CdS photocatalyst. Neither particle size, surface area, nor crystal structure seemed to influence this dark reactivity.[200] Diels–Alder cyclization of diene cation radicals can be induced on irradiated TiO_2 suspensions, although low yields are usually obtained.[99]

Geometric isomerization of substituted olefins[201–204] and cyclopropanes[34a] to attain thermodynamic equilibrium can be effected by irradiation of CdS suspensions, again through the postulated intermediacy of a cation radical. Two mechanistic schemes involving, respectively, competition of a quencher and a substrate for a photoactive site and independent adsorption by the quencher and substrate at independent surface sites.[204] The observation of quantum efficiencies greater than unity implicates a cation radical chain process in these isomerizations.[34a]

A formally symmetry-forbidden [1,3] sigmatropic shift of hydrogen in a bicyclo[4.2.0]octene can be initiated by conducting the reaction on an irradiated CdS surface.[205] This so-far unique reaction suggests that a wealth of unexplored activity mediated by ion radicals may be accessible on irradiated semiconductor suspensions.

Finally, synthetic transformations involving simultaneous oxidation and reductions of adsorbed reactants to form a more complex product have been observed in the synthesis of amino acids from simple gaseous molecules: ammonia, methane, and water.[206–209] Although low efficiencies are reported in this endothermic reaction,[207] its operation implies a mechanism for solar energy conversion, and the possible importance of this observation in prebiotic chemistry is obvious. The efficiency of the conversion could be improved if α-keto acids or glucose were adopted as starting materials.[208] Photoinduced Michael reactions were observed if α,β-unsaturated carbonyl compounds were employed.[208] The photoelectrochemically synthesized amino acids could be converted to oligopeptides by continued irradiation, with different product distributions being observed on TiO_2 and on CdS.[209]

A quite different example of simultaneous oxidation and reduction on an irradiated semiconductor is found in the description of the water-gas shift reaction on TiO_2 [210] and on ZnS.[211] Thus, a number of reactions can be activated by electron or hole trapping on the surface of an irradiated semiconductor, while observing isomerizations or structural reorganization in lieu of net oxidation or reduction chemistry.

4. OPPORTUNITIES AND QUESTIONS IN PHOTOELECTROCHEMISTRY

A complete description of the mechanistic features of chemical reactions on irradiated semiconductor suspensions with the intent of using such knowledge to control selectivity attainable on these photocatalytic surfaces continues to be a most intriguing scientific goal. Given that an at least rudimentary understanding of the course of photoelectrochemical transformations is now in hand, however, it may be useful to identify some other related topics within this field in which pressing scientific questions remain. Such a list is, of course, highly subjective and far from complete. It is offered as a framework from which to view recent developments both in our laboratories and from the several interrelated disciplines that contribute to progress in this area.

4.1. Sensitization

A persistent problem in using semiconductor photocatalysis for solar energy applications is the lack of a photocatalyst that is stable in long-term operation and yet still responsive to large portions of visible wavelengths. The responsiveness of existing metal oxide semiconductors can be expanded by surface attachment, by physisorption, chemisorption, or covalent linkage, with stable dyes absorbing in the desired regions.[212] Furthermore, Grätzel and co-workers have recently shown that highly efficient spectral sensitization of TiO_2 can be attained on porous, rough film surfaces,[213] thus obviating some of the problems previously associated with dye sensitization.[214] A new scheme involving two-photon energy-transfer sensitization has also recently been proposed.[212c] We have shown that organic and inorganic dyes adsorbed on several metal oxides show effective sensitization and long-term stability on metal oxides,[215] and that new routes to porous metal oxides should make possible high optical density sensitization systems.[216]

Even higher local optical densities can be achieved if polymer systems employing pendant dye molecules as light antennae can act as efficient sensitizers.[217] Such systems can be studied by either by photophysical studies of the intermediates generated by sensitization of optically transparent colloidal suspensions[218] or by measuring photocurrent production in an appropriate electrochemical cell.[219] There is also a need for new, highly absorptive chromophores that can be incorporated into these polymers[220] or chemically attached to the relevant photocatalyst.[221] The development of new electrochemical methods to quantitatively characterize adsorption of electro- and photo-active dyes will become increasingly important,[222] as will the development of effective model systems to understand dye interactions within the sensitizing polymer layers.[223]

4.2. New Semiconductors and Supports

Alternatively, the search for new semiconductors with improved photocatalytic properties[224] and expanded light responsiveness might represent another solution to this problem. Our recent discovery that ternary metal oxides, e.g., $PbMoO_4$,[225] may have photocatalytic activity comparable to that seen with TiO_2 represents one possible direction.

An allied goal is to understand how the interaction of existing semiconductors with metal supports[226] or insulating frameworks affects their photocatalytic activities. TiO_2, for example, maintains its activity when supported on silica gel[227] or on glass over which the reactant mixture is circulated.[228] Nafion, a perfluorinated sulfonate cation-exchange membrane, has proved to be a most versatile support, which permits preparation of integrated chemical systems that retain high photoactivity while allowing for reagent compartmentalization.[229] These integrated systems can be studied by luminescence techniques[230] or by surface science methodology.[231]

Supported semiconductors also offer the potential to construct macroscopic devices for vectorial charge transfer. By using a bipolar semiconductor on metal electrode,[232] we have constructed TiO_2 on Pt multipanel arrays on which the photoelectrochemically induced splitting of water occurs with incident solar light sources without external bias.[233] Other metal chalcogenide semiconductors can be employed,[234] and the resulting photocurrent and charge separation can be modeled theoretically.[235]

Observable photoelectrochemistry is attainable only if the photogenerated electron-hole pair can interact within its lifetime with an appropriate donor or acceptor. If the environment of the photogenerated electron-hole pair can be manipulated to prolong the lifetime of the carriers, enhanced photoreactivity will result. One approach to the problem is to transfer the photogenerated electron from the particle on which it was generated by photolysis to another.[236] We have shown, for example, that hydrogen generation can occur when the photoactive semiconductor and the catalyst for gas evolution are on separate particles, presumably through conduction band-mediated electron transfer.[237]

Sustained photochemical activity also requires that the semiconductor itself be protected from corrosion and surface coating by insulating materials. Having shown that aging can be characterized quantitatively by electrochemical methods,[238] we were able to show that an insulating polymer coatings can stabilize the surfaces of colloidal semiconductors[239] and that a conductive polymeric coating allowed retention of photoactivity in powdered catalysts, although with some reduction in efficiency because of light absorption by the conductive polymer.[240] That this stability derives from surface treatment is made probable by the contrasting reactivity seen with TiO_2 prepared under oxidizing and reducing conditions, where a nearly 8-fold enhancement of photocatalytic activity over platinized commercial anatase in the oxidative cleavage of α-methylstyrene was observed.[241] It has also been shown that the photoactivity of TiO_2 (rutile) can be suppressed or enhanced by the surface density of electrons, which in turn is responsive to the pretreatment conditions.[242] Combining the idea of surface protection of the semiconductor with that of separate particles suggests that the preparation of sandwich[243] or layered[244] semiconductors might provide interesting and useful materials.

The order inherent in layered semiconductor can be exploited in the construction of ordered semiconductor materials, i.e., in superlattices or quantum wells.[245] These structures represent ordered arrays of extremely small particles that, while larger than molecular, do not have the spatial extension necessary for the development of complete band structure. As a result, small subcolloidal clusters exhibit unusual quantum mechanical properties. Called q particles,[246] these species have characteristic blue spectral shifts and enhanced reactivity compared with their bulk semiconductor analogs. For example, reductions that are impossible on bulk materials occur with facility on q-particles of PbSe and HgSe.[247] In our work, we have shown that photoactive q-particles can be prepared by ion dilution in Langmuir–Blodgett films[248] or in Nafion.[249]

4.3. Applications

Besides the utility of these systems in achieving chemical control of reactions initiated on surfaces, intensive investigation of these reactions has many practical applications.[250] Among these are their uses as photoimaging systems and in information storage[251] and as techniques for environmental detoxification.[181] Two other applications of semiconductor-based photoelectrochemistry that have been investigated in our research group include the development of a photoelectrochemical detector for high-pressure liquid chromatography[252] and the use of heteropolyoxoanions as soluble analogs for semiconductors to gauge the structural onset of band structure.[253] This class of compounds also shows promise as surface catalysts for inhibiting charge recombination on irradiated semiconductors.[254]

ACKNOWLEDGMENTS

Our research program on characterizing interfacial electron transfer on chemically modified electrode surfaces is sponsored by the U.S. Department of Energy, Office of Basic Energy Sciences, Fundamental Interactions Branch, and that on organic reactive intermediates is supported by the National Science Foundation. Our work on the photochemical decomposition of toxic materials is funded by the U.S. Army Research Office.

REFERENCES

1. Gerischer, H.; Willig, F. *Top. Curr. Chem.* **1976**, *61*, 33.
2. Nozik, A. J. *Ann. Rev. Phys. Chem.* **1978**, *29*, 189.
3. Bard, A. J. *J. Photochem.* **1979**, *10*, 59.
4. Wrighton, M. S. *Acc. Chem. Res.* **1979**, *12*, 303.
5. Gerischer, H. *Pure Appl. Chem.* **1980**, *52*, 2649.
6. Memming, R. *Electrochim. Acta* **1980**, *25*, 77.
7. Bard, A. J. *Science* **1980**, *207*, 139.
8. Heller, A. *Acc. Chem. Res.* **1981**, *14*, 154
9. Bard, A. J. *J. Phys. Chem.* **1982**, *86*, 172.
10. Hodes, G. In *Energy Resources through Photochemistry and Catalysis,* Grätzel, M., Ed. Academic Press, New York, 1983, p. 421.
11. (a) Aspnes, D. E.; Heller, A. *J. Phys. Chem.* **1983**, *87*, 4919; (b) Heller, A. *Science* **1984**, *223*, 1141.
12. Gerischer, H. *J. Phys. Chem.* **1984**, *88*, 6096.
13. Hodes, G.; Grätzel, M. *Nouv. J. Chim.* **1984**, *8*, 509.
14. Grätzel, M. *Heterogeneous Photochemical Electron Transfer.* CRC Press, Boca Raton, FL, 1988.
15. Schiavello, M. *NATO Adv. Sci. Ser.* **1988**, *C237*, 351.
16. Pichat, P. *NATO Adv. Sci. Ser.* **1986**, *C174*, 533.
17. Pichat, P. In *Photocatalysis,* Serpone, N.; Pelizzetti, E., Eds. Wiley, New York, 1989, Chapter 8, p. 217.
18. Mozzanega, H.; Herrmann, J. M.; Pichat, P. *J. Phys. Chem.* **1979**, *83*, 2251.
19. Pichat, P.; Herrmann, J. M.; Disdier, J.; Courbon, H.; Mozzanega, M. N. *Nouv. J. Chim.* **1981**, *5*, 627.
20. (a) Fox, M. A.; Ogawa, H.; Muzyka, J. *Proc. Electrochem. Soc.* **1988**, *88–14*, 9; (b) Fox, M. A.; Ogawa, H. *J. Inform. Record. Mater.* **1990**, *5/6*, 351.
21. Fox, M. A.; Ogawa, H.; Pichat, P. *J. Org. Chem.* **1989**, *54*, 3847.
22. Hussein, F. H.; Pattenden, G.; Rudham, R.; Russell, J. J. *Tetrahedron Lett.* **1984**, 3363.
23. Muzyka, J. L.; Fox, M. A. *J. Org. Chem.* **1990**, *55*, 209.
24. (a) Bahnemann, D. W.; Hoffmann, M. R. *Proc. Electrochem. Soc.* **1988**, *88–14*, 74; (b) Bahnemann, D. W.; Moenig, J.; Chapman, R. *J. Phys. Chem.* **1987**, *91*, 3782.
25. Krauetler, B.; Bard, A. J. *J. Am. Chem. Soc.* **1978**, *100*, 5985.
26. Jaeger, C. D.; Bard, A. J. *J. Phys. Chem.* **1979**, *83*, 3146.
27. Dean, M. H.; Stimming, U. *Proc. Electrochem. Soc.* **1988**, *88–14*, 460.
28. Findlayson, M. F.; Wheeler, B. L.; Kakuta, N.; Park, K.; Bard, A. J.; Campion, A.; Fox, M. A.; Webber, S. E.; White, J. M. *J. Phys. Chem.* **1985**, *89*, 5676.
29. Kabir-ud-Din; Owen, R. C.; Fox, M. A. *J. Phys. Chem.* **1981**, *85*, 1679.
30. Tomkiewicz, M.; Shen, W. M. *Proc. Electrochem. Soc.* **1988**, *88–14*, 468.
31. Kanno, T.; Oguchi, T.; Sakuragi, H.; Tokumaru, K. *Tetrahedron Lett.* **1980**, *21*, 467.
32. Schwerzel, R. E. *Proc. Electrochem. Soc.* **1983**, *83*, 513.

33. (a) Harada, H.; Sakata, T.; Ueda, T. *J. Am. Chem. Soc.* **1985**, *107*, 1773; (b) Harada, H.; Ueda, T. Sakata, T. *J. Phys. Chem.* **1989**, *93*, 1542; (c) Harada, H.; Sakata, T.; Ueda, T. *Sixth Int. Conf. Storage Solar Energy* **1986**, Abstr. C-37; (d) Harada, H.; Sakata, T.; Ueda, T. *Proc. Electrochem. Soc.* **1988**, *88–14*, 144.
34. (a) Carson, P. A.; de Mayo, P. *Can. J. Chem.* **1987**, *65*, 976; (b) Baird, N. C.; Draper, A. M.; de Mayo, P. *Can. J. Chem.* **1988**, *66*, 1579.
35. Hetterich, W.; Kisch, H. *Chem. Ber.* **1988**, *121*, 15.
36. Kamioka, H.; Suzuki, M.; Tamiya, E.; Karube, I. *J. Mol. Catal.* **1989**, *54*, 1.
37. Kim, Y. S.; Fox, M. A. Unpublished results.
38. (a) Krauetler, B.; Bard, A. J. *J. Am. Chem. Soc.* **1978**, *100*, 2339; (b) Yoneyama, H.; Takao, Y.; Tamura, H.; Bard, A. J. *J. Phys. Chem.* **1983**, *87*, 1417.
39. For example, Radlick, P.; Klem, R.; Spurlock, S.; Sims, J. J.; van Tamelen, E. E.; Whitesides, T. *Tetrahedron Lett.* **1968**, 5117.
40. Dunn, W. W.; Aikawa, Y.; Bard, A. J. *J. Am. Chem. Soc.* **1981**, *103*, 3456.
41. (a) Kölle, U.; Moser, J.; Grätzel, M. *Inorg. Chem.* **1985**, *24*, 2253; (b) Nakabayashi, S.; Fujishima, A.; Honda, K. *J. Am. Chem. Soc.* **1985**, *107*, 250; (c) Gerischer, H. *J. Phys. Chem.* **1984**, *88*, 6096.
42. Howe, R.; Grätzel, M *J. Phys. Chem.* **1985**, *89*, 4495; (b) Serwicka, E. *Colloid. Surf.* **1985**, *13*, 287; (c) Amorelli, A.; Evanc, J. C.; Rowlands, C. C. *J. Chem. Soc., Faraday Trans. I* **1988**, *84*, 1723.
43. Rothenberger, G.; Moser, J.; Grätzel, M.; Serpone, N.; Sharma, D. K. *J. Am. Chem. Soc.* **1985**, *107*, 8054.
44. Borgarello, E.; Kiwi, J.; Pelizzetti, E.; Visca, M.; Grätzel, M. *J. Am. Chem. Soc.* **1981**, *103*, 6324.
45. Nosaka, Y.; Fox, M. A. *J. Phys. Chem.* **1988**, *92*, 1893.
46. Nosaka, Y.; Fox, M. A. *J. Phys. Chem.* **1986**, *90*, 6521.
47. Fox, M. A., Lindig, B.; Chen, C. C. *J. Am. Chem. Soc.* **1982**, *104*, 5828.
48. Serpone, N.; Pelizzetti, E. *NATO Adv. Sci. Ser.* **1986**, *C174*, 51.
49. (a) Warman, J. M.; de Haas, M. P.; Grätzel, M.; Infelta, P. P. *Nature (London)* **1984**, *310*, 305; (b) Kunst, M.; Beck, G.; Tributsch, H. *J. Electrochem. Soc.* **1984**, *131*, 954.
50. Kamat, P. V.; Fox, M. A. *J. Phys. Chem.* **1983**, *87*, 59.
51. Kiwiet, N. J., Ph.D. Dissertation, University of Texas, 1989.
52. Kajiwara, T.; Hashimoto, K.; Kawai, T.; Sakata, T. *J. Phys. Chem.* **1982**, *86*, 4516.
53. Draper, R. B.; Fox, M. A. *J. Phys. Chem.* **1990**, *94*, 4628.
54. Albery, W. J.; Bartlett, P. N.; Wilde, C. P.; Darwent, J. R. *J. Am. Chem. Soc.* **1985**, *107*, 1854.
55. Draper, R. B.; Fox, M. A. *Langmuir* **1990**, *6*, 1396.
56. Draper, R. B.; Fox, M. A.; Pelizzetti, E.; Serpone, N. *J. Phys. Chem.* **1989**, *93*, 1938.
57. Serpone, N.; Al-Ekabi, H.; Patterson, B.; Pelizzetti, E.; Minero, C.; Pramauro, E.; Fox, M. A.; Draper, B. *Langmuir* **1989**, *5*, 250.
58. Anpo, M.; Tomonari, M.; Fox, M. A. *J. Phys. Chem.* **1989**, *93*, 7300.
59. Tributsch, H. *NATO Adv. Sci. Ser.* **1988**, *C 237*, 297.
60. (a) Amouyal, E. *NATO Adv. Sci. Ser.* **1986**, *C 174*, 253; (b) Haruvy, Y., Rafaeloff, R.; Rajbenbach, L. A. *NATO Adv. Sci. Ser.* **1986**, *C 174*, 267.

61. (a) Kiwi, J.; Kalyanasundaram, K.; Grätzel, M. *Struct. Bond.* **1981**, *49*, 37; (b) Kiwi, J. *Proc. Electrochem. Soc.* **1988**, *88-14*, 275.
62. Palmisano, L.; Augugliaro, V.; Sclafani, A.; Schiavello, M. *J. Phys. Chem.* **1988**, *92*, 6710.
63. (a) Kalyanasundaram, K. *Solar Cells* **1985**, *15*, 93; (b) Kalyanasundaram, K. In *Energy Resources through Photochemistry and Catalysis*, Grätzel, M., Ed. Academic Press, New York, 1983.
64. (a) Möllers, F.; Tolle, H. J.; Memming, R. *J. Electrochem. Soc.* **1974**, *121*, 1160; (b) Krauetler, B.; Bard, A. J. *J. Am. Chem. Soc.* **1978**, *100*, 4317; (c) Kobayashi, T., Taniguchi, Y.; Yoneyama, H.; Tamara, H. *J. Phys. Chem.* **1983**, *87*, 768; (d) Ulmann, M.; Augustynski, J. *Proc. Electrochem. Soc.* **1988**, *88-14*, 176; (e) Nakato, Y.; Ueda, K.; Tsubomua, H. *Proc. Electrochem. Soc.* **1988**, *88-14*, 190.
65. Lauermann, I.; Meissner, D.; Memming, R. *Proc. Electrochem. Soc.* **1988**, *88-14* 190.
66. (a) Pichat, P. *New J. Chem.* **1987**, *11*, 129; (b) Kawai, T.; Sakata, T. *Nature (London)* **1980**, *286*, 474.
67. Persaud, L.; Bard, A. J.; Campion, A.; Fox, M. A.; Mallouk, T. E.; Webber, S. E.; White, J. M. *Inorg. Chem.* **1987**, *26*, 3825.
68. Fox, M. A.; Pettit, T. L. *Langmuir* **1989**, *5*, 1056.
69. Persaud, L.; Bard, A. J.; Campion, A.; Fox, M. A.; Mallouk, T. E.; Webber, S. E.; White, J. M. *J. Am. Chem. Soc.* **1987**, *109*, 7309.
70. (a) Porter, J. D.; Heller, A.; Aspnes, D. E. *Nature (London)* **1985**, *313*, 664; (b) Allongue, P.; Souteyrand, E.; Allemand, L.; Cachet, H. *Proc. Electrochem Soc.* **1988**, *88-14*, 212.
71. Kakuta, N.; Park, K. H.; Finlayson, M.; Bard, A. J.; Campion, A.; Fox, M. A., White, J. M.; Webber, S. E.; *J. Phys. Chem.* **1985**, *89*, 732.
72. Sobczynski, A.; Yildiz, A.; Bard, A. J.; Campion, A.; Fox, M. A.; Mallouk, T. E.; Webber, S. E.; White, J. M. *J. Phys. Chem.* **1988**, *92*, 2311.
73. Sobczynski, A.; Bard, A. J.; Campion, A., Fox, M. A.; Mallouk, T. E.; Webber, S. E.; White, J. M. *J. Phys. Chem.* **1989**, *93*, 401.
74. Fox, M. A. *NATO Adv. Studies Ser.* **1988**, *C 237*, 445.
75. Fox, M. A. *Top. Curr. Chem.* **1987**, *142*, 72.
76. Fox, M. A. *Chim. Ind. (Milan)* **1986**, *68*, 59.
77. Fox, M. A. *NATO Adv. Sci. Inst. Ser.* **1986**, *174*, 363
78. Fox, M. A.; Chen, C. C.; Park, K.; Younathan, J. N. *ACS Symp. Ser.* **1985**, *278*, 69.
79. Fox, M. A. *Top. Org. Electrochem.* **1985**, *4*, 177.
80. Fox, M. A. *Acct. Chem. Res.* **1983**, *16*, 314.
81. Fox, M. A.; Pichat, P. In *Photoinduced Electron Transfer*, Fox, M. A.; Chanon, M., Eds. Elsevier, Amsterdam, 1988, Vol. D, Chapter 6.1, p. 241.
82. Fox, M. A. In *Photocatalysis,* Serpone, N.; Pelizzetti, E., Eds. Wiley, New York, 1989, Chapter 13, p. 421.
83. Norris, J. R., Jr.; Meisel, D., Eds. *Photochemical Energy Conversion*. Elsevier, Amsterdam, 1989.
84. Wrighton, M. S., Ed. *Interfacial Photoprocesses: Energy Conversion and Synthesis*. American Chemical Society, Washington, D.C., 1980.

85. Nozik, A. J. *Photoeffects at Semiconductor–Electrolyte Interfaces.* American Chemical Society, Washington, D.C., 1981.
86. Santhanam, K. S. V; Sharon, M., Eds. *Photoelectrochemical Solar Cells.* Elsevier, Amsterdarn, 1988.
87. Ginley, D. S.; Nozik, A.; Armstrong, N.; Honda, K.; Fujishima, A., Sakata, T., Eds. *Photoelectrochemistry and Electrosynthesis on Semiconducting Materials.* The Electrochemical Society, Pennington, NJ, 1988.
88. Kalyanasundaram, K.; Grätzel, M.; Pelizzetti, E. *Coord. Chem. Rev.* **1986,** *69,* 57.
89. Harriman, A.; West, M. A., Eds. *Photogeneration of Hydrogen. Academic Press, New York, 1982.*
90. Fox, M. A. In *Photoinduced Electron Transfer,* Fox, M. A.; Chanon, M., Eds. Elsevier, Amsterdam, 1988, Vol. D, Chapter 5.1, p. 1.
91. Izumi, I.; Dunn, W. W.; Wilbourn, K. O.; Fan, F. R. F.; Bard, A. J. *J. Phys. Chem.* **1980,** *84,* 3207.
92. Izumi, I.; Fan, F. R. F.; Bard, A. J. *J. Phys. Chem.* **1981,** *85,* 218.
93. (a) Jaeger, C. D.; Bard, A. J. *J. Phys. Chem.* **1979,** *83,* 3146; (b) Anpo, M.; Shima, T.; Kubokawa, Y. *Chem. Lett.* **1985,** 1799.
94. Cunningham, J.; Srijaranai, S. *J. Photochem. Photobiol.* **1988,** *A 43,* 329.
95. Fujihira, M.; Satoh, Y.; Osa, T. *Bull. Chem. Soc. Jpn.* **1982,** *55,* 666.
96. Sackett, D. D.; Fox, M. A. *J. Phys. Org. Chem.* **1988,** *1,* 103.
97. Al-Ekabi, H.; Draper, A. M.; de Mayo, P. *Can. J. Chem.* **1989,** *67,* 1061.
98. Ohtani, B.; Okugawa, Y.; Nishimoto, S.; Kagiya, T. *J. Phys. Chem.* **1987,** *91,* 3550.
99. Muzyka, J.; Fox, M. A. Unpublished results.
100. Kawai, T.; Sakata, T. *Chem. Lett.* **1981,** 81.
101. (a) Krasnovskii, A.; Nikandrov, V. V. *Priroda (Moscow)* **1988,** *12,* 39; (b) Schrauzer, G. N.; Guth, T. D.; Salehi, J.; Strampach, N.; Hui, L. N.; Palmer, M. R. *NATO Adv. Sci. Ser.* **1986,** *174,* 509.
102. Kawai, T.; Sakata, T. *Chem. Commun.* **1980,** 694.
103. Kawai, T.; Sakata, T. *Nature (London)* **1980,** *286,* 474.
104. Pichat, P.; Herrmann, J. M.; Disdier, J.; Courbon, H.; Mozzanega, M. N. *Nouv. J. Chim.* **1981,** *5,* 627.
105. Pichat, P.; Mozzanega, M. N.; Disdier, J.; Herrmann, J. M. *Nouv. J. Chim.* **1982,** *6,* 559.
106. Pichat, P.; Herrmann, J. M.; Courbon, H.; Disdier, J.; Mozzanega, M. N. *Can. J. Chem. Eng.* **1982,** *60,* 27.
107. Enea, O. *Electrochim. Acta* **1986,** *31,* 405.
108. Enea, O.; Bard, A. J. *Nouv. J. Chim.* **1985,** *9,* 361.
109. Sakata, T.; Kawai, T. *Chem. Phys. Lett.* **1981,** *80,* 341.
110. Ichou, I. A.; Formenti, M.; Teichner, S. J. *Stud. Surf. Sci. Catal.* **1984,** *19,* 297.
111. Teratani, S.; Naimichi, J.; Taya, K.; Tanaki, K. *Bull. Chem. Soc. Jpn.* **1982,** *55,* 1688.
112. Carlson, T.; Griffin, G. L. *J. Phys. Chem.* **1986,** *90,* 5896.
113. Naito, S. *Chem. Commun.* **1985,** 1211.
114. Domen, K.; Naito, S.; Ohnishi, T.; Tamaru, K. *Chem. Lett.* **1982,** 555.
115. Prahov, L. T.; Disidier, J.; Herrmann, J.; Pichat, P. *Int. J. Hydrogen Ener.* **1984,** *9,* 397.

116. Fraser, I. M.; MacCallum, J. R. *J. Chem. Soc., Faraday Trans. I* **1986**, *82*, 2747.
117. Fraser, I. M.; MacCallum, J. R. *J. Chem. Soc., Faraday Trans. I* **1986**, *82*, 607.
118. Nishimoto, S.; Ohtani, B.; Shirai, H.; Kagiya, T. J. *Chem. Soc., Faraday Trans. II* **1986**, 661.
119. Ward, M. D.; Razdil, J. F.; Grasselli, R. K. *J. Phys. Chem.* **1984**, *88*, 4210.
120. Teichner, S. J.; Formenti, M. *NATO Adv. Sci. Ser.* **1985**, *C 146*, 379.
121. Matthews, R. W. *J. Catal.* **1988**, *111*, 264.
122. Walker, A.; Formenti, M.; Meriaudeau, P.; Teichner, S. J. *J. Catal.* **1977**, *50*, 237.
123. Cunningham, J.; Hodnett, B. K. *J. Chem. Soc., Faraday Trans. I* **1981**, *77*, 2777.
124. Blake, N. R.; Griffin, G. L. *J. Phys. Chem.* **1988**, *92*, 5697.
125. Ohtani, B.; Kakimoto, M.; Miyadzu, H.; Nishimoto, S.; Kagiya, T. *J. Phys. Chem.* **1988**, *92*, 5773.
126. Muñuera, G.; González-Elipe, A. R.; Fernández, A.; Malet, P.; Espinós, J. P. *J. Chem. Soc. Faraday I* **1989**, *85*, 1279.
127. Amorelli, A.; Evans, J. C.; Rowlands, C. C. *J. Chem. Soc. Faraday I* **1988**, *84*, 1723.
128. Yamagata, S.; Baba, R.; Fujishima, A. *Bull. Chem. Soc. Jpn.* **1989**, *62*, 1004.
129. Sakamaki, K.; Matsunaga, S.; Itoh, K.; Fujishima, A.; Gohshi, Y. *Surf. Sci.* **1989**, *219*, L531.
130. (a) Okamoto, K.; Yamamoto, Y.; Tanaka, H.; Tanaka, M.; Itaya, A. *Bull Chem. Soc. Jpn.* **1985**, *58*, 2015; (b) Kawaguchi, H. *Environ. Technol. Lett.* **1984**, *5*, 471.
131. (a) Peral, J.; Casado, J.; Domenech, J. *J. Photochem. Photobiol.* **1988**, *A 44*, 209; (b) Peral, J.; Casado, J.; Domenech, J. *Electochim. Acta* **1989**, 1327.
132. Bonhomme, G.; Lemaire, J. *C. R. Seances Acad. Sci, Ser. II* **1986**, *302*, 769.
133. El-Akabi, H.; Serpone, N. *J. Phys. Chem.* **1988**, *92*, 5726.
134. (a) Fox, M. A.; Chen, C. C. *J. Am. Chem. Soc.* **1981**, *103*, 6757; (b) Fox, M. A.; Chen, C. C. *J. Photochem.* **1981**, *17*, 119.
135. (a) Kanno, T.; Oguchi, T.; Sakuragi, H.; Tokumaru, K. *Tetrahedron Lett.* **1980**, *21*, 467; (b) Al-Ekabi, H.; de Mayo, P. *J. Org. Chem.* **1987**, *52*, 4756.
136. San Filippo, J.; Chern, C. I.; Valentine, J. S. *J. Org. Chem.* **1976**, *41*, 1077.
137. Spada, L. T.; Foote, C. S. *J. Am. Chem. Soc.* **1980**, *102*, 393.
138. Schenck, G. O.; Schulte-Ellte, K. *Annalen* **1958**, *107*, 1773.
139. Fox, M. A.; Chen, C. C. *Tetrahedron Lett.* **1983**, *24*, 547.
140. Oae, S.; Watanabe, Y.; Fujimori, K. *Tetrahedron Lett.* **1982**, *1189*.
141. (a) Nelsen, S. F. *J. Am. Chem. Soc.* **1981**, *103*, 2096; (b) Nelsen, S. F.; Kapp, D. L.; Akaba, R.; Evans, D. H. *J. Am. Chem. Soc.* **1986**, *108*, 6863.
142. Fox, M. A.; Chen, C. C. *J. Comput. Chem.* **1983**, *4*, 488.
143. Anpo, M.; Matsuura, T., Eds. *Photochemistry on Solid Surfaces.* Elsevier, Amsterdam, 1989.
144. Anpo, M.; Nishiguchi, H.; Fujii, T. *Res. Chem. Intermed.* **1990**, in press.
145. Seri-Levy, A.; Samuel, J., Farin, D.; Avnir, D. *Stud. Surf. Sci. Catal.* **1989**, *47*, 353.
146. Tsuchiya, M.; Ebbesen, T. W.; Nishimura, Y.; Sakuragi, H.; Tokumaru, K. *Stud. Org. Chem.* **1988**, *33*, 133.
147. Pichat, P.; Herrmann, J. M.; Disdier, J.; Mozzanega, M. N. *J. Phys. Chem.* **1979**, *83*, 3122,

148. Ward, M. D.; Brazdi, J. F., Jr.; Grasselli, R. K. U.S. Patent 4,571,290, 1986: *Chem. Abstr.* **1986**, *105*, 162141.
149. Anpo, M.; Yabuta, M.; Kodama, S.; Kubokawa, Y. *Bull. Chem. Soc. Jpn.* **1986**, *59*, 259.
150. Kodama, S.; Yagi, S. *J. Phys. Chem.* **1989**, *93*, 4833.
151. Fujihira, M.; Satoh, Y.; Osa, T. *J. Electroanal. Chem.* **1981**, *126*, 277. (b) Pichat, P.; Disdier, J.; Herrmann, J. M.; Vaudano, P. *Nouv. J. Chim.* **1986**, *10*, 545; (c) Liang, J. J.; Liu, T. J. *J. Chin. Chem. Soc.* **1986**, *33*, 133; (d) Matthews, R. W. *Aust. J. Chem.* **1987**, *40*, 667.
152. (a) Sakata, T.; Kawai, T. *Sympos. Org. Chem. Jpn.* **1981**, *39*, 589; (b) Djehri, N.; Formenti, M.; Juillet, F.; Teichner, S. J. *Discuss. Faraday Soc.* **1975**, *58*, 184; (c) Courbon, H.; Herrmann, J. M.; Pichat, P. *J. Catal.* **1981**, 72, 129.
153. Giannotti, C. G.; LeGreneur, S.; Watts, O. *Tetrahedron Lett.* **1983**, *24*, 5071.
154. Fox, M. A.; Sackett, D. D.; Younathan, J. N. *Tetrahedron* **1987**, 22.
155. Younathan, J. N. Ph.D. Dissertation, University of Texas, 1985.
156. (a) Barton, D. H. R.; Haynes, R. K.; LeClerc, G.; Magnus, P. D.; Menzies, I. D. *J. Chem. Soc., Perkin Trans. I* **1975**, 2055; (b) Tang, R.; Yue, H. J.; Wolf, J. F.; Mares, F. *J. Am. Chem. Soc.* **1979**, *100*, 5248; (c) Landis, M. E.; Madoux, D. C. *J. Am. Chem. Soc.* **1979**, *101*, 5106; (d) Haynes, R. K.; Probert, M. K.; Wilmot, I. D. *Aust. J. Chem.* **1978**, *31*, 1737; (e) Haynes, R. K. *Aust. J. Chem.* **1978**, *31*, 121, 131.
157. Mozzangea, M. N.; Herrmann, J. M.; Pichat, P. *Tetrahedron Lett.* **1977**, *2965*.
158. Fujihira, M.; Satoh, Y.; Osa, T. *Nature (London)* **1981**, *293*, 206.
159. Fox, M. A.; Chen, C. C.; Younathan, J. N. *J. Org. Chem.* **1984**, *49*, 1969.
160. For example, see (a) Cunningham, J.; Meriaudeau, P. *J. Chem. Soc., Faraday Trans. I* **1976**, 1499; (b) Cundall, R. B.; Rudham, R.; Salim, M. S. *J. Chem. Soc., Faraday Trans. I* **1976**, 1642; (c) Teratani, S.; Nakamichi, J; Taya, K.; Tanaka, K. *Bull. Chem. Soc. Jpn.* **1982**, *55*, 1688; (d) Nishimoto, S.; Ohtani, B.; Kagiya, T. *J. Chem. Soc., Faraday Trans. I* **1985**, *81*, 2467; (e) Shinoda, T.; Ohkawa, K. *Kagaku Giijutsu* **1985**, *80*, 89; (f) Li, Q.; Naito, S.; Tamaru, K. *Xiamen Daxue Xuebao* **1986**, *25*, 63; *Chem. Abstr.* **1986**, *105*, 231659; (g) Pichat, P.; Mozzanega, M. N.; Courbon, H. *J. Chem. Soc., Faraday Trans. I* **1987**, *83*, 697; (h) Hussein, F. H.; Rudham, R. *J. Chem. Soc., Faraday Trans. I* **1987**, *83*, 1631; (i) Korzhak, A. V.; Kuchmii, S. Ya.; Kryukov, A. I. *Teor. Eksp. Khim.* **1987**, *23*, 181; (j) Ohno, M.; Uzawa, H.; Miyazaki, T.; Tarama, K. *Chem. Lett.* **1987**, 779.
161. (a) Palmisano, L.; Sclafani, A.; Schiavello, M.; Augugliaro, V.; Coluccia, S.; Marchese, L. *New J. Chem.* **1988**, *12*, 137; (b) Cuendet, P.; Grätzel, M. *J. Phys. Chem.* **1987**, *91*, 654.
162. Domenech, J.; Peral, J. *J. Chem. Res. (S)* **1987**, *360*.
163. Furlong, D. N.; Wells, D.; Sasse, W. H. F. *Aust. J. Chem.* **1986**, *39*, 757.
164. Pincock, J. A.; Pincock, A. L.; Fox, M. A. *Tetrahedron* **1985**, *41*, 4107.
165. (a) Bücheler, J.; Zeug, N.; Kisch, H. *J. Am. Chem Soc.* **1985**, *107*, 1459; (b) Zeug, N.; Bücheler, J.; Kisch, H. *Angew. Chem. Int. Ed. Engl.* **1982**, *21*, 77; (c) Yanagida, S.; Azuma, T.; Sakurai, H. *Chem. Lett.* **1982**, 1069; (d) Yanagida, S.; Azuma, T.; Midori, Y., Pac, C.; Sakurai, H. *J. Chem. Soc., Perkin Trans. II*, **1985**, 1487.
166. Chen, M. J.; Fox, M. A. *J. Am. Chem. Soc.* **1983**, *105*, 4497.

167. (a) Ohtani, B.; Osaki, H.; Nishimoto, N.; Tagiya, T. *J. Am. Chem. Soc.* **1986**, *108*, 308; (b) Nishimoto, S.; Ohtani, B.; Yoshikawa, T.; Kagiya, T. *J. Am. Chem. Soc.* **1983**, *105*, 7180.
168. Ohtani, B.; Osaki, H.; Nishimoto, S.; Kagiya, T. *Tetrahedron Lett.* **1986**, *27*, 2019.
169. Fox, M. A.; Younathan, J. N. *Tetrahedron* **1986**, *42*, 6285.
170. Pavlik, J. W.; Tantayanon, S. *J. Am. Chem. Soc.* **1981**, *103*, 6755.
171. (a) Hema, M. A.; Ramakrishnan, V.; Kuriacose, J. C. *Indian J. Chem.* **1978**, *16*, 619; (b) Kasturirangan, H.; Ramakrishnan, V.; Kuriacose, J. C. *J. Catal.* **1981**, *69*, 216.
172. Mozzanega, M. N.; Herrmann, J. M.; Pichat, P. *Chem. Phys. Lett.* **1980**, *74*, 523.
173. Abdul-Wahab, A. A.; Dulay, M.; Fox, M. A.; Kim, Y.-S. *Catal. Lett.* **1990**, *5*, 369.
174. (a) Hoyer, W.; Asmus, K. D. *Sixth International Conf. Photochem. Conver. Solar Energy*, Abstr 55, 1986; (b) Spikes, J. D. *Photochem. Photobiol.* **1981**, *34*, 549.
175. Harada, K.; Hisanaga, T.; Tanaka, K. *New J. Chem.* **1987**, *11*, 597.
176. Grätzel, C. K.; Jirousek, M.; Grätzel, M. *J. Mol. Catal.* **1987**, *39*, 347; (b) Harada, K.; Hisanaga, T.; Tanaka, K. *Nouv. J. Chim.* **1987**, *11*, 597.
177. Hsiao, C. Y.; Lee, C. L.; Ollis, D. F. *J. Catal.* **1983**, *82*, 418.
178. (a) Barbeni, M.; Pramauro, E.; Pelizzetti, E.; Borgarello, E.; Serpone, N. *Chemosphere* **1985**, *14*, 195; (b) Al-Ekabi, H.; Serpone, N. *J. Phys. Chem.* **1988**, *92*, 5726; (c) Matthews, R. W. *J. Catal.* **1986**, *97*, 565; (d) Ollis, D.F. *J. Catal.* **1986**, *97*, 569.
179. Pelizzetti, E.; Borgarello, M.; Minero, C.; Pramauro, E.; Borgarello, E.; Serpone, N. *Chemosphere* **1988**, *17*, 499.
180. Matthews, R. W. *J. Catal.* **1988**, *113*, 549.
181. (a) Schiavello, M., Ed. *Photocatalysis and Environment, Trends and Applications.* Kluwer Academic Publishers, Dordrecht, 1988; (b) Cesareo, D.; di Domenico, A.; Marchini, S.; Passerini, L.; Tosato, M. L. *NATO Adv. Sci Ser.* **1986**, *A174*, 593; (c) Oliver, B. G.; Carey, J. H. *NATO Adv. Sci Ser.* **1986**, *A174*, 630; (d) Ollis, D. F. *NATO Adv. Sci Ser.* **1986**, *A174*, 651.
182. Fox, M. A.; Pettit, T. L. *J. Org. Chem.* **1985**, *50*, 5013.
183. Kiwi, J. *J. Phys. Chem.* **1986**, *90*, 1493.
184. Suzuki, S.; Horiuchi, N.; Hori, Y. *Kankyo Kagaku* **1984**, *9*, 24; *Chem. Abstr.* **1984**, *102*, 35786.
185. Nosaka, Y.; Ishizuka, Y.; Norimatsu, K.; Miyama, H. *Bull. Chem. Soc. Jpn.* **1984**, 3066.
186. Cai, Z. S.; Kuntz, R.R. *Langmuir* **1988**, *4*, 830.
187. Darwent, J. R.; Lepre, A. *J. Chem. Soc., Faraday Trans. II* **1986**, *82*, 1457.
188. Fan, I. J.; Chien, Y. C. *Sci. Sinica* **1978**, *21*, 663.
189. (a) Frank, A. J.; Goren, Z.; Willner, I. *Chem. Commun.* **1985**, 1029; (b) Anpo, M.; Shima, T.; Kodama, S.; Kubokawa, Y. *J. Phys. Chem.* **1987**, *91*, 4305; (c) Baba, R.; Nakabayashi, S.; Fujishima, A.; Honda, K. *J. Am. Chem Soc.* **1983**, *109*, 2273.
190. (a) Yamataka, H; Sato, N.; Ichihara, J.; Hanafusa, T.; Teratani, S. *Chem. Commun.* **1985**, 788; (b) Chen, Y.; Li, W.; Dong, Y. *Cuihua Xuebao* **1985**, *6*, 91: *Chem. Abstr.* **1985**, *102*, 206507; (c) Boonstra, A.H.; Mutsaers, C. A. *J. Phys. Chem.* **1975**, *79*, 2025; (d) Yun, C.; Anpo, M.; Kodama, S.; Kubokawa, Y. *Chem. Commun.* **1980**, 609.
191. Kamat, P. V. *J. Photochem. Photobiol.* **1985**, *28*, 513.

192. (a) Brown, G. T.; Darwent, J. R. *J. Phys. Chem.* **1984**, *88*, 4955; (a) Brown, G. T.; Darwent, J. R. *J. Chem. Soc., Faraday Trans. I* **1984**, *80*, 1631.
193. Yanagida, S.; Ishimaru, Y.; Miyake, Y.; Shiragami. T.; Pac, C.; Hashimoto, K.; Sakata, T. *J. Phys. Chem.* **1989**, *93*, 2576.
194. (a) Okada, K.; Hisamitsu, K.; Mukai, T. *Chem Commun.* **1980**, 941; (b) Okada, K.; Hisamitsu, K.; Takahashi, Y.; Hanaoka, T.; Miyashi, T.; Mukai, T. *Tetrahedron Lett.* **1984**, *25*, 5311.
195. Ikezawa, H.; Kutal, C. *J. Org. Chem.* **1987**, *52*, 3299.
196. Lahiry, S.; Haldar, C. *Solar Energy* **1986**, *37*, 71.
197. Al-Ekabi, H.; de Mayo, P. *J. Phys. Chem.* **1986**, *90*, 4075.
198. Barber, R. A.; de Mayo, P.; Okada, K. *Chem. Commun.* **1982**, 1073.
199. Draper, A. M.; Ilyas, M.; de Mayo, P.; Ramamurthy, V. *J. Am. Chem. Soc.* **1984**, *106*, 6222.
200. Ilyas, M.; de Mayo, P. *J. Am. Chem. Soc.* **1985**, *107*, 5093.
201. Al-Ekabi, H.; de Mayo, P. *Chem. Commun.* **1984**, 1231.
202. Al-Ekabi, H.; de Mayo, P. *J. Phys. Chem.* **1985**, *89*, 5815.
203. Hasegawa, T.; de Mayo, P. *Chem. Commun.* **1985**, 1534.
204. Hasegawa, T.; de Mayo, P. *Langmuir* **1986**, *2*, 362.
205. de Mayo, P.; Wenska, G. *Tetrahedron* **1987**, *43*, 1661.
206. Dunn, W. W.; Aikawa, Y.; Bard, A. J. *J. Am. Chem. Soc.* **1981**, *103*, 6893.
207. Reiche, H.; Bard, A. J. *J. Am. Chem. Soc.* **1979**, *101*, 2239.
208. Sakata, T. *J. Photochem. Photobiol.* **1985**, *29*, 205.
209. Onoe, J.; Kawai, T.; Kawai, S. *Chem. Lett.* **1985**, 1667.
210. Fang, S. M.; White, J. M. *J. Phys. Chem.* **1982**, *86*, 3126.
211. Kisch, H.; Schlamann, W. *Chem. Ber.* **1986**, *119*, 3483.
212. (a) Gerischer H. *Photochem. Photobiol.* **1972**, *16*, 243; (b) Gerischer, H; Willig, F. *Top. Curr. Chem.* **1976**, *61*, 31; (c) Iyoda, T.; Sakamaki, K.; Shimidizu, T.; Honda, K. *Proc. Electrochem. Soc.* **1988**, *88-14*, 169.
213. (a) Desilvestro, J.; Grätzel, M.; Kavan, L.; Moser, J.; Augustynski, J. *J. Am. Chem. Soc.* **1985**, *107*, 2988; (b) Vlachopoulos, N.; Liska, P.; Augustiynski, J.; Grätzel, M. *J. Am. Chem. Soc.* **1988**, *110*, 1216.
214. (a) Memming, R. *Photochem. Photobiol.* **1972**, *16*, 325; (b) Ryan, M. A.; Spitler, M. T. *Langmuir* **1988**, *4*, 861; (c) Kirk, A. D.; Langford, C. H.; Joly, C. S.; Lesage, R.; Sharma, D. K. *Chem. Commun.* **1984**, 961; (d) Sonntag, L. P.; Spitler, M. T. *J. Phys. Chem.* **1985**, *89*, 1453; (e) Liang, Y.; Goncalves, A. M. P. *J. Phys. Chem.* **1985**, *89*, 3290.
215. Dabestani, R.; Bard, A. J.; Campion, A.; Fox, M. A.; Mallouk, T. E.; Webber, S. E.; White, J. M. *J. Phys. Chem.* **1988**, *92*, 1872.
216. Kudo, A.; Steinberg, M.; Bard, A. J.; Campion, A.; Fox, M. A.; Mallouk, T. E.; Webber, S. E.; White, J. M. *J. Electrochem. Soc.* **1990**, *137*, 3846.
217. (a) Fox, M. A.; Britt, P.F. *Macromol.* **1990**, *23*, 4533; (b) Kamat, P. V.; Fox, M. A. *J. Phys. Chem.* **1984**, *88*, 2297; (c) Kamat, P. V.; Fox, M. A. *J. Photochem.* **1984**, *24*, 285; (c) Kamat, P. V.; Fox, M. A. *J. Electroanal. Interfac. Electrochem.* **1983**, *159*, 49.
218. Kamat, P. V.; Fox, M. A. *Chem. Phys. Lett.* **1983**, *102*, 379.
219. Kamat, P. V.; Fox, M. A. *J. Electrochem. Soc.* **1984**, *131*, 1032; (b) Fox, M. A.; Kabir-ud-Din, *J. Phys. Chem.* **1979**, *83*, 1800.

220. (a) Fox, M. A.; Kamat, P. V.; Fatiadi, A. J. *J. Am. Chem. Soc.* **1984**, *106*, 1191; (b) Fox, M. A.; Owen, R. *Am. Chem. Soc. Symp. Ser.* **1981**, *146*, 337; (c) Fox, M. A.; Owen, R. *J. Am. Chem. Soc.* **1980**, *102*, 6559.
221. Fox, M. A.; Hohman, J. R.; Kamat, P. V. *Can. J. Chem.* **1983**, *61*, 888; (b) Hohman, J. R.; Fox, M. A.; *J. Am. Chem. Soc.* **1982**, *104*, 401; (c) Fox, M. A.; Nobs, F.; Voynick, T. *J. Am. Chem. Soc.* **1980**, *102*, 4029.
222. Creager, S. E.; Fox, M. A. *Electroanal. Chem.* **1989**, *258*, 431.
223. (a) Fox, M. A.; Britt, P. F. *J. Phys. Chem.* **1990**, *94*, 6351; (b) Fox, M. A., Britt, P. F. *Photochem. Photobiol.* **1990**, *51*, 129.
224. Anpo, M. *Proc. Electrochem. Soc.* **1988**, *88-14*, 34.
225. Kudo, A.; Steinberg, M.; Bard, A. J.; Campion, A.; Fox, M. A.; Mallouk, T. E.; Webber, S. E.; White, J. M. *Catal. Lett.* **1990**, *5*, 61.
226. Kudo, A.; Steinberg, M.; Bard, A. J.; Campion, A.; Fox, M. A.; Mallouk, T. E.; Webber, S. E.; White, J. M. *J. Catal.* **1990**, in press.
227. Ueno, A.; Kakuta, N.; Park, K. H.; Finlayson, M. F.; Bard, A. J.; Campion, A.; Fox, M. A.; Mallouk, T. E.; Webber, S. E.; White, J. M. *J. Phys. Chem.* **1985**, *89*, 3828.
228. Serpone, N.; Borgarello, E.; Harris, R.; Cahill, P.; Borgarello, E. *Sol. Energy Mater.* **1986**, *14*, 121.
229. (a) Krishnan, M.; White, J. R.; Fox, M. A.; Bard, A. J. *J. Am. Chem. Soc.* **1983**, *105*, 7002; (b) Mau, A. W. H.; Huang, C. B.; Kakuta, N.; Park, K. H.; Bard, A. J.; Campion, A.; Fox, M. A.; Webber, S. E.; White, J. M. *J. Am. Chem. Soc.* **1984**, *106*, 6537; (c) Dabestani, R.; Wang, X., Bard, A. J.; Campion, A.; Fox, M. A.; Webber, S. E.; White, J. M. *J. Phys. Chem.* **1986**, *90*, 2729; (d) Fox, M. A. *Pure Appl. Chem.* **1988**, *60*, 1013.
230. Finlayson, M.; Park, K. H.; Kakuta, N.; Bard, A. J.; Campion, A.; Fox, M. A.; Mallouk, T. E.; Webber, S. E.; White, J. M. *J. Luminescence* **1988**, *39*, 205.
231. (a) Kakuta, N.; Park, K. H.; Finlayson, M. F.; Bard, A. J.; Campion, A.; Fox, M. A.; Webber, S. E.; White, J. M. *J. Phys. Chem.* **1985**, *89*, 5028; (b) Kakuta, N.; Park, K. H.; Finlayson, M. F.; Bard, A. J.; Campion, A.; Fox, M. A.; Webber, S. E.; White, J. M. *Surf. Interface Anal.* **1985**, *7*, 295; (c) Kakuta, N.; Park, K. H.; Finlayson, M. F.; Bard, A. J.; Campion, A.; Fox, M. A.; Webber, S. E.; White, J. M.; Finlayson, M. *Am. Chem. Soc. Sympos. Ser.* **1985**, *288*, 566; (d) Kakuta, N.; Park, K. H.; Finlayson, M. F.; Bard, A. J.; Campion, A.; Fox, M. A.; Webber, S. E.; White, J. M.; Finlayson, M. *ACS Petrol. Prepr.* **1984**, *29*, 867; (e) Kakuta, N.; Park, K. H.; Finlayson, M. F.; Bard, A. J.; Campion, A.; Fox, M. A.; Webber, S. E.; White, J. M.; Finlayson, M. *J. Phys. Chem.* **1985**, *89*, 48.
232. Fox, M. A. *Nouv. J. Chim.* **1987**, *11*, 129.
233. Smotkin, E.; Bard, A. J.; Campion, A.; Fox, M. A.; Mallouk, T. E.; Webber, S. E.; White, J. M. *J. Phys. Chem.* **1986**, *90*, 4604.
234. Smotkin, E.; Cervera-March, S.; Bard, A. J.; Campion, A.; Fox, M. A.; Mallouk, T. E.; Webber, S. E.; White, J. M. *J. Phys. Chem.* **1987**, *91*, 6.
235. Smotkin, E.; Cervera-March, S.; Bard, A. J.; Campion, A.; Fox, M. A.; Mallouk, T. E.; Webber, S. E.; White, J. M. *J. Electrochem. Soc.* **1986**, *135*, 567.
236. Serpone, N.; Borgarello, E.; Grätzel, M. *Chem. Commun.* **1985**, 318; (c) Pichat, P.; Disdier, J.; Herrmann, J. M.; Mozzangea, M. N.; Hoang-Van, C. *Proc.*

Electrochem. Soc. **1988**, *88-14*, 50; (d) Spanhel, L.; Henglein, A.; Weller, H. *Ber. Bunsenges. Phys. Chem.* **1987**, *91*, 1359.
237. Sobczynski, A.; Bard, A. J.; Campion, A.; Fox, M. A.; Mallouk, T. E.; Webber, S. E.; White, J. M. *J. Phys. Chem.* **1987**, *91*, 3316.
238. Kiwiet, N. J.; Fox, M. A. *J. Electrochem. Soc.* **1990**, *137*, 561.
239. Nosaka, Y.; Fox, M. A. *Langmuir* **1987**, *3*, 1147.
240. Yildiz, A.; Sobczynski, A.; Bard, A. J.; Campion, A.; Fox, M. A.; Mallouk, T. E.; Webber, S. E.; White, J. M. *Langmuir* **1989**, *5*, 148.
241. Becker, W. G.; Truong, M. M.; Ai, C. C.; Hamel, N. N. *J. Phys. Chem.* **1989**, *93*, 4882.
242. Heller, A.; Degani, Y.; Johnson, D. W., Jr.; Gallagher, P. K. *Proc. Electrochem. Soc.* **1988**, *88-14*, 23.
243. Rabani, J. *J. Phys. Chem.* **1989**, *93*, 7707.
244. Micic, O. I.; Nenadovic, M. T.; Peterson, M. W.; Nozik, A. J. *J. Phys. Chem.* **1987**, *91*, 1295.
245. Nozik, A. J.; Turner, J. A.; Peterson, M. W. *J. Phys. Chem.* **1988**, *92*, 2493.
246. (a) Henglein, A. *Prog. Colloid Polym. Sci.* **1987**, 731; (b) Henglein, A. *Top. Curr. Chem.* **1988**, *143*, 113; (c) Brus, L. *NATO Adv. Sci. Ser.* **1986**, *C 174*, 111; (d) Micic, O. I.; Nenadovic, M. T.; Rajh, T.; Dimitrijevic, N. M.; Nozik, A. J. *NATO Adv. Sci. Ser.* **1986**, *C 174*, 213.
247. Nedeljkovic, J. M.; Nenadovic, M. T.; Mici, O. T.; Nozik, A. J. *J. Phys. Chem.* **1986**, *90*, 12.
248. Smotkin, E. S.; Lee, C.; Bard, A. J.; Campion, A.; Fox, M. A.; Mallouk, T. E.; Webber, S. E.; White, J. M. *Chem. Phys. Lett.* **1988**, *152*, 265.
249. Smotkin, E. S.; Rabenberg, L. K.; Salomon, K.; Bard, A. J.; Campion, A.; Fox, M. A.; Mallouk, T. E.; Webber, S. E.; White, J. M. *J. Phys. Chem.* **1990**, *94*, 7543.
250. Fox, M. A.; Chanon, M., Eds. *Photoinduced Electron Transfer. Volume D. Applications*. Elsevier, Amsterdam, 1988.
251. Alfimov, M. V.; Sazhnikov, V. A. in *Photoinduced Electron Transfer. Volume D. Applications*, Fox, M. A.; Chanon, M., Eds. Elsevier, Amsterdam, 1988, p. 474.
252. Fox, M. A.; Tien, T. P. *Anal. Chem.* **1988**, *60*, 2278.
253. Fox, M. A.; Cardona, R.; Gaillard, E. *J. Am. Chem. Soc.* **1987**, *107*, 6347.
254. Keita, B.; Nadjo, L. *J. Chim. Phys. Phys.-Chim. Biol.* **1988**, *85*, 223.

THERMAL AND PHOTOCHEMICAL ACTIVATION OF AROMATIC DONORS BY ELECTRON TRANSFER

Christian Amatore and Jay K. Kochi

1. Introduction 56
2. Reversible Oxidation Potentials of Arenes 58
   2.1. Electrochemical Measurements in Trifluoroacetic Acid 59
   2.2. Electrochemical Measurements in Acetonitrile 59
   2.3. Structural Correlations of the Arene Oxidation Potentials . . . 62
   2.4. Solvation of Arene Cation Radicals 63
3. Electron-Transfer Activation of Arenes 65
   3.1. Kinetics of Oxidative Substitution of Methylarenes with Iron(III) Oxidants 65
   3.2. Mechanistic Formulation of Electron Transfer from Arenes . . 68
   3.3. Application of Marcus Theory to Arene Activation by Electron Transfer 72
4. Side Chain Substitution of Methylarenes by Electron Transfer: The Deprotonation of Cation Radical Intermediates 78
   4.1. Rates of Deprotonation of Methylarene Cation Radicals . . . 79
   4.2. Direct Observation of the Kinetic Acidities of Arene Cation Radicals 81

Advances in Electron Transfer Chemistry,
Volume 1, pages 55–148.

ISBN: 1-55938-167-1

4.3. Free Energy Relationship for the Deprotonation of Methylarene Cation Radicals 83
5. Side Chain Versus Nuclear Substitution of Methylarene Cation Radicals 89
6. Direct Observation of Electrophilic Aromatic Substitution by Electron Transfer: Photoactivation of Nuclear Nitration Via Charge-Transfer Complexes 97
6.1. Charge-Transfer Nitration of Aromatic Donors 98
6.2. Dimethoxybenzenes as the Aromatic Donors in Nitration . . . 100
6.3. Haloanisoles as the Aromatic Donors in Nitration 103
6.4. Structural Variation in the Kinetics of Ion Radical Pair Collapse 107
6.5. Decay Kinetics of Ion Radical Pairs as Reactive Intermediates 108
6.6. Comments on the Mechanism of Aromatic Nitration 113
7. Ion Pair Versus Radical Pair Annihilation of Arene Cation Radicals: Salt and Solvent Effects 115
8. Electron Transfer Activation in the Thermal and Photochemical Osmylations of Aromatic EDA Complexes with Osmium(VIII) Tetroxide 128
8.1. Aromatic EDA Complexes with Osmium(VIII) Tetroxide . . . 131
8.2. Thermal Osmylation of Naphthalene, Anthracene, and Phenanthrene 131
8.3. Charge-Transfer Osmylation of Benzene, Naphthalene, and Anthracene 133
8.4. Time-Resolved Spectra of Arene Cation Radicals in Charge-Transfer Osmylation 134
8.5. Common Features in Thermal and Charge-Transfer Osmylations 135
8.6. Electron Transfer in the Charge-Transfer Osmylation of Arenes 137
8.7. Electron Transfer as the Common Theme in Arene Osmylation 140
9. Epilogue 141
Acknowledgments 143
References 143

1. INTRODUCTION

Activation of arenes by electron transfer was initially formulated by Dewar and co-workers,[1] who showed that oxidative substitution of an electron-rich aromatic donor by manganic acetate in Eq. 1 (ArH = *p*-methoxytoluene) followed first-order kinetics for each reactant. This together with the observation of a substantial kinetic isotope effect for the deuterated donor p-$MeOC_6H_4CD_3$ and an inverse dependence on added $Mn(OAc)_2$ led to the proposal that the side chain substitution in Eq. 1 occurred via an initial electron transfer, i.e.

$$ArH + 2Mn(OAc)_3 \rightarrow ArOAc + 2Mn(OAc)_2 + HOAc \qquad (1)$$

$$ArH + Mn(III) \underset{k_{-1}}{\overset{k_1}{\rightleftharpoons}} ArH^{\cdot+} + Mn(II) \qquad (2)$$

$$ArH^{\cdot+} \xrightarrow{k_2} Ar^{\cdot} + H^+ \qquad (3)$$

$$Ar^{\cdot} + Mn(OAc)_3 \rightarrow ArOAc + Mn(OAc)_2 \qquad (4)$$

Scheme 1.

where Ar = p-$CH_3OC_6H_4CH_2$. An analogous electron-transfer mechanism is also applicable to the *nuclear* substitution of benzene to yield phenyl esters with the aid of such strong oxidants as cobaltic trifluoroacetate (Eq. 5).[2]

$$C_6H_6 + 2Co(O_2CCF_3)_3 \rightarrow$$

$$C_6H_5O_2CCF_3 + 2Co(O_2CCF_3)_2 + CF_3CO_2H \qquad (5)$$

In both cases, the cation radical ($ArH^{\cdot+}$) is the reactive intermediate derived by electron transfer from the aromatic donor to the oxidant (compare Eq. 2).[3] As such, the reduction potential E^0_{red} of the oxidant and the oxidation potential E^0_{ox} of the arene are important factors in determining the viability of both processes for aromatic substitution.[4] Although the one-electron electrode potentials of inorganic oxidants are readily determined, the corresponding values for arenes are generally unavailable owing to the transient character of most aromatic cation radicals.[5] Accordingly it is necessary to establish reliable values of E^0_{ox} for various aromatic donors in order to delineate the mechanism of the electron-transfer step, particularly with regard to the dependence of the rate (log k_1) with the driving force $\Delta G^0 = -\mathcal{F}(E^0_{red} - E^0_{ox})$ where $\mathcal{F}$ is the Faraday constant.[6] Since the latter is usually endergonic (i.e., $\Delta G^0 >> 0$) for most arene/oxidant pairs, the back electron transfer (k_{-1}) is likely to be facile and lead to a reversible electron transfer.[7] Under these conditions, the follow-up reaction (Eq. 3) could represent the rate-limiting step. Depending on the relative magnitudes of the rate constants, the electron-transfer mechanism in Scheme 1 may be either easy to verify experimentally (i.e., when $k_2 >> k_{-1}$) or difficult to distinguish (i.e., when $k_2 << k_{-1}$) from a concerted mechanism in which the electron transfer is simultaneous with bond scission (or formation).[6] The latter also bears on the more general question as to the participation of electron-transfer mechanisms in various electrophilic processes. For example, in the nitration of arenes with NO_2^+, the activation process is conventionally formulated as a direct addition of the electrophile to form the critical Wheland intermediate, i.e.[8]

$$ArH + NO_2^+ \xrightarrow{k_1} Ar\begin{smallmatrix} H^+ \\ NO_2 \end{smallmatrix}, \text{ etc.} \tag{6}$$

Alternatively, the same electrophilic substitution can be formulated as a two-step mechanism in which electron-transfer activation is followed by the subsequent collapse of the ion radical pair, i.e.[9]

$$ArH + NO_2^+ \overset{k_1}{\rightleftharpoons} [ArH^{\cdot+}, NO_2^{\cdot}] \xrightarrow{\text{fast}} Ar\begin{smallmatrix} H^+ \\ NO_2 \end{smallmatrix}, \text{ etc.} \tag{7}$$

The experimental distinction between these mechanistic formulations is dependent on whether the ion radical pair is (or is not) formed. However, the independent proof of the ion radical pair has not been forthcoming because of its expectedly transitory character. Thus the arene ion radical pair is not formed in sufficient concentrations to observe in a thermal (adiabatic) process, since its rate of annihilation will always be faster than its rate of production.[10] An alternative nonadiabatic (photochemical) approach is clearly required to test the viability of the electron-transfer mechanism.[11] To address this question, we believe it is necessary to lay the groundwork for the various mechanistic aspects of aromatic activation by electron transfer. To this end a few selected examples will be treated rigorously in this chapter to illustrate each facet of the problem. For a wider coverage of aromatic chemistry the reader is referred to general monographs.[12]

2. REVERSIBLE OXIDATION POTENTIALS OF ARENES

Electrochemical methods offer the most direct access to the oxidation potentials of arenes; among the readily available techniques, cyclic voltammetry (CV) is the simplest and the most convenient to use, particularly in organic solvents. Unfortunately with the exception of highly condensed polycyclic and very electron-rich systems, the cyclic voltammograms of most aromatic compounds exhibit irreversible behavior at sweep rates < 100 V s^{-1}.[13] This is shown by the absence of the cathodic component on the return potential sweep, largely the result of competition from fast follow-up reactions of the metastable arene cation radicals. However the recent development of microvoltammetric electrodes has allowed cyclic voltammograms to be recorded at sweep rates exceeding 100,000 V s^{-1}.[14] Such high scan rates are not compatible with conventional electrodes as a result of the pronounced distortion of the voltammograms associated with large uncompensated ohmic drops and long time constants of the capacitive phenomena. We thus resorted to ultramicroelectrodes consisting of either a gold or platinum wire (diameter 10 μm) imbedded in a glass insulator so that only a cross section of the disc was exposed to the solution.[15]

2.1. Electrochemical Measurements in Trifluoroacetic Acid

Arene cation radicals are known to be significantly more persistent in acidic media such as sulfuric and trifluoroacetic acids.[16] Accordingly the application of the microvoltammetric technique to various arenes was carried out in trifluoroacetic acid with gold microelectrodes.[15] The optimum sweep rates of 200–2000 V s^{-1} to obtain chemically reversible cyclic voltammograms are included in Figure 1, and the values of the reversible potential E^0_{Ar} for the homologous series from toluene to hexa-

$$ArCH_3 \underset{}{\overset{E^0_{Ar}}{\rightleftharpoons}} ArCH_3^{+\cdot} \tag{8}$$

methylbenzene are listed in Table I. The table also includes the vertical ionization potentials (*IP*) of the same series of methylarenes from their He(I) photoelectron spectra.[17]

2.2. Electrochemical Measurements in Acetonitrile

Arene cation radicals are significantly less persistent in the more basic solvent acetonitrile, and cyclic voltammetry in this solvent necessitated the use of significantly enhanced sweep rates to achieve the degree of reversibility obtained in the more acidic trifluoroacetic acid. The greater ohmic drop under these conditions was partially offset by an increased concentration of supporting electrolyte.[18] The validity of the oxidation potentials of E^0_{ox} = 1.58, 1.69, 1.75, and 1.77 V vs SCE for hexamethylbenzene (HMB), pentamethylbenzene (PMB), durene (DUR), and 1,2,3,4-tetramethylbenzene (TMB), respectively, was established by

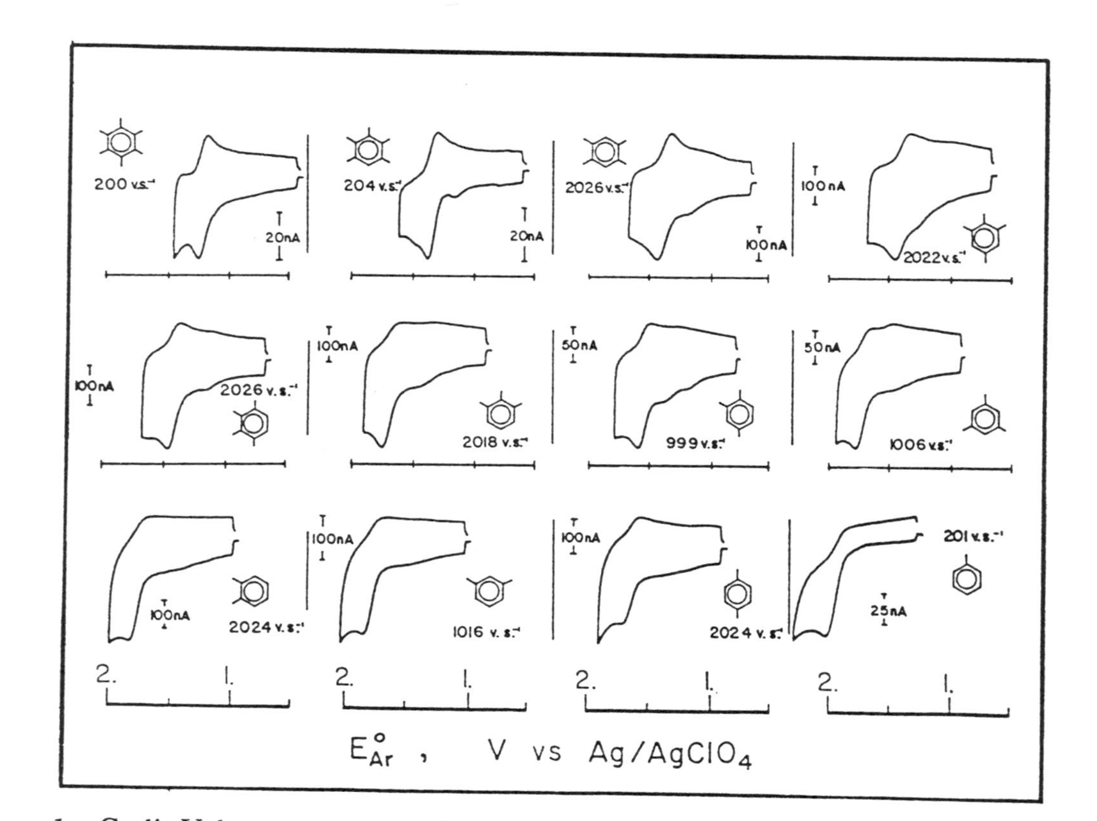

Figure 1. Cyclic Voltammograms at a Gold Microelectrode of Methylarenes in Trifluoroacetic Acid Containing 0.1 *M* tetra-*n*-Butylammonium Perchlorate (TBAP) at the Sweep Rates Indicated.

Table I. Ionization Potentials and Reversible Oxidation Potentials of Methylarenes.

$ArCH_3^{\bullet+}$	I_p[a]	E^0_{Ar}[b]	$ArCH_3^{\bullet+}$	I_p[a]	E^0_{Ar}[b]
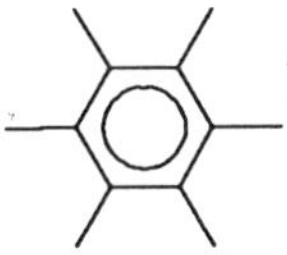	7.85(7.83)[c]	1.62(1.58)	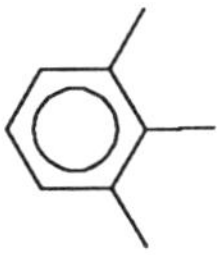	8.42	1.99
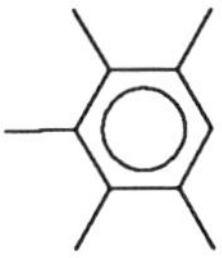	7.92	1.75(1.69)	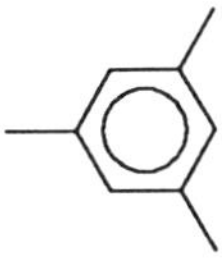	8.40	2.11
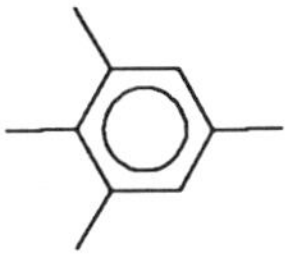	8 06	1.83	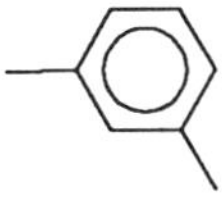	8.56	2.14
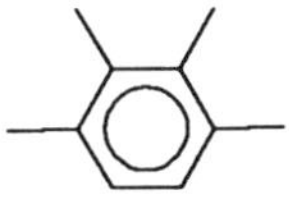	8.14	1.82(1.77)	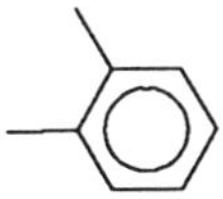	8.56	2.13
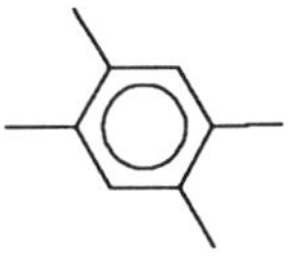	8.03	1.83(1.75)		8.44	2.06
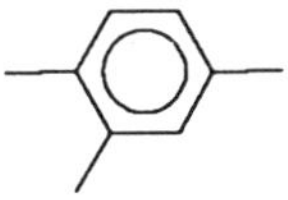	8.27	1.89	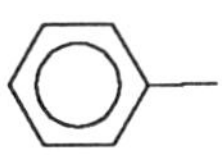	8.76	2.40

[a]From photoelectron spectra in electron volts.

[b]In trifluoroacetic acid containing 7% vol trifluoroacetic anhydride and 0.1 *M* tetra-*n*-butylammonium perchlorate at 25°C. Potentials in volts vs SCE.

[c]Values in parentheses measured in acetonitrile containing 0.6 *M* tetra-*n*-butylammonium tetrafluoroborate.

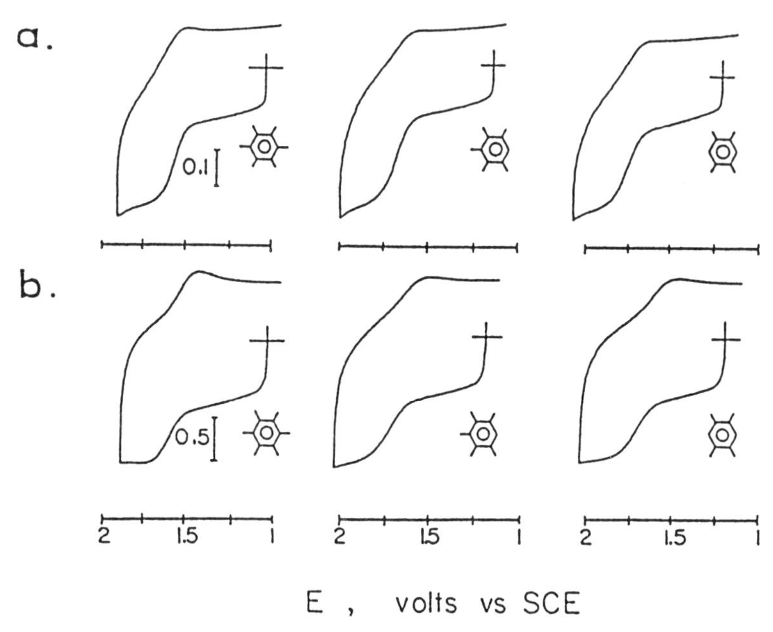

Figure 2. Cyclic Voltammograms of Methylarenes in Acetonitrile, 0.6 *M* NBu_4BF_4 at 20°C. Platinum Ultramicroelectrode (Diameter 10 μm). Scan Rate: (a) 4900, (b) 38,800 Vs^{-1}. (Current Scales are Given in μA.) $ArCH_3$ Concentration: 1 m*M*.

their consistency with scan rate variation, as shown by the comparisons of the cyclic voltammograms in Figure 2.

2.3. Structural Correlations of the Arene Oxidation Potentials

The examination of the cyclic voltammograms in Figures 1 and 2 shows that the values of E^0_{Ar} progressively decrease with increasing numbers of methyl substituents[20]—the difference between hexamethylbenzene and toluene being more than 700 mV. Furthermore, the chemical reversibility of the cyclic voltammograms (as indicated by the ratios of the cathodic and anodic peak currents i^c_p/i^a_p at a given scan rate), generally parallels the magnitude of E^0_{Ar}, being the most reversible for the highly methylated benzenes. The correlation of the reversible oxidation potentials of the methylarenes in trifluoroacetic acid solution with the ioniza-

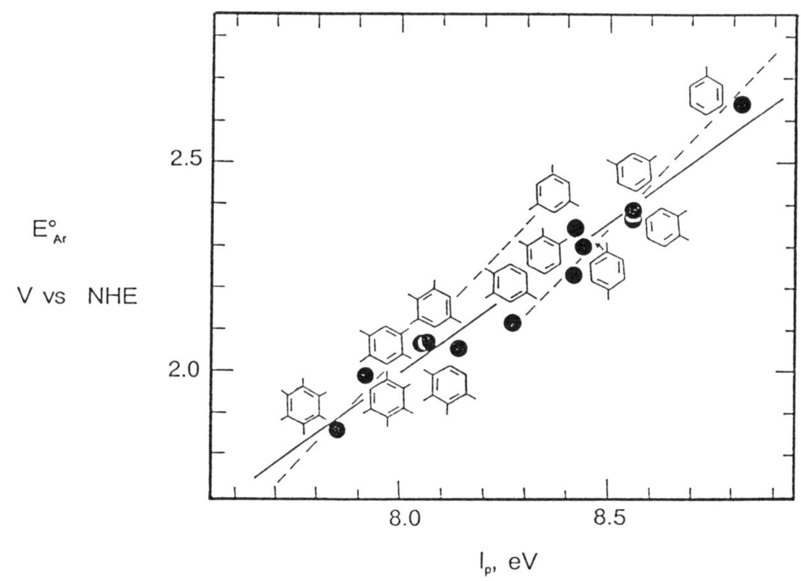

Figure 3. Correlation of the Reversible Oxidation Potentials and the Vertical Ionization Potentials of the Methylarenes Identified in Table I.

tion potentials in the gas phase is shown in Figure 3. The line in the graph describes the relationship

$$E^0_{Ar} = 0.71\,IP - 3.67 \tag{9}$$

when E^0_{Ar} is given in volts vs the $Ag/AgClO_4$ standard electrode and IP is given in electron volts.

2.4. Solvation of Arene Cation Radicals

The slope of considerably less than unity indicates that the energetics of the gas phase ionization are not completely mirrored in the solution oxidation. In particular, the contribution from solvation is implicit in the values of E^0_{Ar} whereas it is not in the values of IP. The difference in the free energy (ΔG^0_s) associated with solvation changes can be expressed as

$$\Delta G^0_s = (G^0_{Ar^+})_s - (G^0_{Ar})_s - (G^0_{Ar^+})_g + (G^0_{Ar})_g \tag{10}$$

where the subscripts g and s refer to the gas phase and solvation states. Correspondingly the standard oxidation potential is

$$E^0_{Ar} = (1/\mathcal{F})\{\Delta G^0_s + (G^0_{Ar^+})_g - (G^0_{Ar})_g\} + C \quad (11)$$

where $\mathcal{F}$ is the Faraday constant and C is a constant determined by the particular working electrode and reference electrode. Under the same circumstances, the vertical ionization potential is

$$IP = (1/\mathcal{F})\{(G^0_{Ar^{+\bullet}})_g - (G^0_{Ar})_g\} \quad (12)$$

where $(G_{Ar^{+\bullet}})_g$ is the free energy of formation of the arene cation radical in the unrelaxed state in which it has the same nuclear coordinates as those in the neutral arene. The combination of Eqs. 10 and 11 yields

$$E^0_{Ar} = IP + (\Delta G^0_r + \Delta G^0_s) / \mathcal{F} + C \quad (13)$$

where ΔG^0_r, the reorganization energy of $ArH^{\bullet+}$, represents the difference $(G^0_{Ar^+})_g - (G^0_{Ar^{+\bullet}})_g$. Comparison of Eq. 13 with the experimental relationship in Eq. 9 indicates that $(\Delta G^0_r + \Delta G^0_s)$ cannot be considered as a constant term for the series of arenes under consideration. Instead it varies with IP or E^0_{Ar}, as given by the combination of Eqs. 9 and 13, i.e.

$$\Delta G^0_r + \Delta G^0_s = -0.29\mathcal{F}\, IP + C' \quad (14)$$

where $C' = (4.1 + C)$. Such a relationship implies that $(\Delta G^0_r + \Delta G^0_s)$ decreases by about 10 kcal mol^{-1} in covering the gamut of aromatic hydrocarbons from hexamethylbenzene to toluene at the extremes.[21] Interestingly this trend also parallels the structural changes of increasing size of the arene moiety resulting from poly-methyl substitution. Since size is an important factor in solvation energies, the deviation of the slope in Figure 3 from unity may well represent variations of mainly ΔG^0_s. For example, a closer inspection of the data suggests that the energy correlation can be (arbitrarily) dissected into three classes of arene donors that are differentiated according to the degree of steric encumbrance by methyl groups that will hinder access to solvation of the cation radical in the ring plane. If so, the most hindered methylarenes (I = hexamethylbenzene, pentamethylbenzene, and durene) and the least hindered methylarenes (II = toluene, xylenes, and mesitylene) both follow an energy correlation with unit slope (see dashed lines in Figure 3), given by Eqs. 15 and 16, respectively.[22]

$$E^0_{Ar}(\text{I}) = IP - 5.87 \tag{15}$$

$$E^0_{Ar}(\text{II}) = IP - 6.08 \tag{16}$$

The potential difference of 0.21 V (~ 5 kcal mol^{-1}) represents the difference in the solvation energy between these two series of arenes. By comparison, a rough estimate based on the Born model (using an average radius of 3.5 Å for the alkylbenzenes and the dielectric constant of 39.5 for trifluoroacetic acid)[23] indicates a solvation energy of ~ 24 kcal mol^{-1} for the methylarene cation radical. Thus the relatively small variation of ~20% between the two series can be readily accommodated by a corresponding difference of the equivalent radius.[24] Within this context, ΔG^0_s can be considered as a constant for each series of arene cation radicals.

3. ELECTRON-TRANSFER ACTIVATION OF ARENES

A series of tris(phenanthroline)iron complexes FeL_3^{3+} can be exploited for the study of electron transfer from aromatic hydrocarbons. The outer-sphere iron(III) oxidants FeL_3^{3+} are particularly suited for mechanistic studies since they are well-behaved in solution to allow for meaningful kinetics (i.e., they are substitution inert and their reversible potentials E^0_{Fe} can be varied systematically by nuclear substitution of the

$$\text{Fe(X-phen)}_3^{2+} \underset{}{\overset{E^0_{Fe}}{\rightleftharpoons}} \text{Fe(X-phen)}_3^{3+} \tag{17}$$

phenanthroline ligand (e.g., for X-phen with X = H, 5-Cl, and 5-NO_2, E_{Fe} = 1.09, 1.19, and 1.29 V vs SCE, respectively).[25]

3.1. Kinetics of Oxidative Substitution of Methylarenes with Iron(III) Oxidants

When a solution of tris(phenanthroline)iron(III) is mixed with hexamethylbenzene in either trifluoroacetic acid or acetonitrile, there is a color change from blue to red diagnostic of the reduction to iron(II). Side chain substitution leads to pentamethylbenzyl trifluoroacetate and acetamide, which can be isolated in essentially quantitative yields in CF_3COOH and CH_3CN, respectively. Furthermore in anhydrous acetonitrile containing pyridine, the *N*-benzylpyridinium salt is obtained with the following stoichiometry:

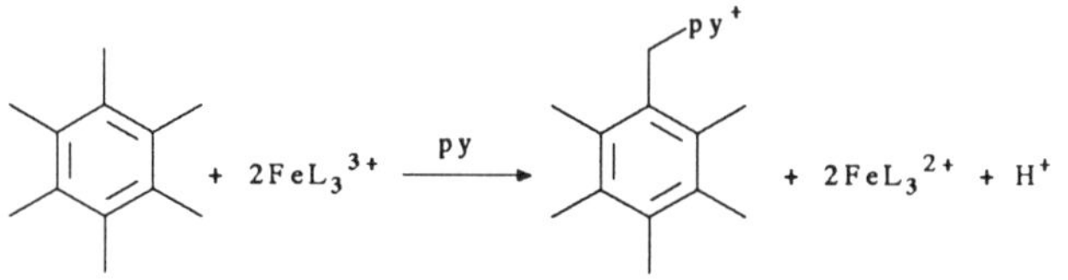

$$\text{HMB} + 2\text{FeL}_3^{3+} \xrightarrow{\text{py}} \text{ArCH}_2\text{py}^+ + 2\text{FeL}_3^{2+} + \text{H}^+ \qquad (18)$$

The kinetics of the oxidative substitution in Eq. 18 were followed by measuring either the disappearance of FeL_3^{3+} at 650 nm (ε = 540 M^{-1} cm^{-1}) or the appearance of FeL_3^{2+} at 510 nm ($\varepsilon = 1.1 \times 10^4 M^{-1}$ cm^{-1}). In the presence of excess hexamethylbenzene (HMB), the rate varied strongly with the amount of pyridine added. The dependence on pyridine derives from the general mechanism in Scheme 1 for which the deprotonation step (Eq. 3) is forced to be irreversible by mass action. Accordingly, the mechanism for iron(III) reduction becomes

$$\text{ArCH}_3 + \text{FeL}_3^{3+} \underset{k_{-1}}{\overset{k_1}{\rightleftharpoons}} \text{ArCH}_3^{\cdot+} + \text{FeL}_3^{2+} \qquad (19)$$

$$\text{ArCH}_3^{\cdot+} + \text{py} \xrightarrow{k_2} \text{ArCH}_2^{\cdot} + \text{Hpy}^+ \qquad (20)$$

$$\text{ArCH}_2^{\cdot} + \text{FeL}_3^{3+} \xrightarrow{\text{fast}} \text{FeL}_3^{2+} + \text{ArCH}_2^{+}\text{, etc.} \qquad (21)$$

Scheme 2.

Since the oxidation of benzyl radicals in Eq. 21 is rapid,[26] the consideration of the steady-state behaviors of $[ArCH_3^{\cdot+}]$ and $[ArCH_2\bullet]$ in Scheme 2 leads to the rate law for the disappearance of iron(III) as

$$1 + \left\{\frac{k_{-1}[\text{Fe(III)}]_0}{k_2(\text{py})}\right\}\ln\frac{[\text{Fe(III)}]}{[\text{Fe(III)}]_0} + \left\{\frac{k_{-1}[\text{Fe(III)}]_0}{k_2(\text{py})}\right\}\left\{1-\frac{[\text{Fe(III)}]}{[\text{Fe(III)}]_0}\right\} = -2k_1[\text{ArCH}_3]_0 t \qquad (22)$$

where $[Fe(III)]_0$ and $[Fe(III)]$ represent the concentrations of tris-phenanthrolineiron(III) initially and at time t, respectively, and $[ArCH_3]_0$ is the concentration of methylarene in > 10-fold excess.

The general rate expression in Eq. 22 thus contains three rate constants that are relevant to this study, namely, k_1 and k_{-1} for electron

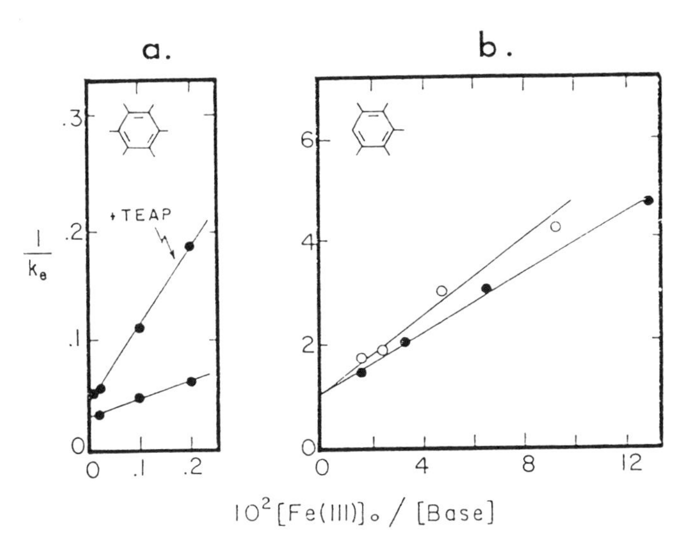

Figure 4. Dependence of the Experimental Rate Constant k_e for the Oxidative Substitution of (*a*) Hexamethylbenzene and (*b*) Pentamethylbenzene by FeL_3^{3+} on the Concentration of Pyridine (•) and 2,6-Lutidine (o). The Effect of 0.1 *M* Tetramethylammonium Perchlorate is Shown in (*a*).

transfer and k_2 for proton transfer. Let us now consider how each of these rate constants can be extracted from the experimental data.

3.1.1. *The Electron-Transfer Rate Constant* k_1

This is obtained at sufficiently high pyridine concentrations such that $k_2[\mathrm{py}]$ is fast relative to back electron transfer, $k_{-1}[\mathrm{Fe(III)}]_0$. Under these conditions, the rate of iron(III) disappearance obeys first-order kinetics, i.e.,

$$\ln \frac{[\mathrm{Fe(III)}]}{[\mathrm{Fe(III)}]_0} = -2k_e t \tag{23}$$

and it follows from Eqs. 22 and 23 that

$$\frac{1}{k_e} = \frac{1}{k_1} + \frac{k_{-1}}{k_1 k_2}\frac{[\mathrm{Fe(III)}]_0}{[\mathrm{py}]} \tag{24}$$

A typical experimental dependence of this rate constant on the iron(III) and pyridine is shown in Figure 4, in which the intercept yields the rate constant k_1 directly. The values of k_1 determined in this way for the oxidation of various methylarenes by several substituted phenanthroline-iron(III) oxidants are listed in Table II. Since the electron transfer rate constant listed for the first entry in the table was subject to a small negative salt effect, all kinetic measurements were carried out at constant ionic strength with either 0.1 *M* tetraethylammonium or lithium perchlorate. Two criteria were used to evaluate the validity of this measure of the electron transfer rate constants. First, the value of k_1 in Table II was within experimental error found to be the same for the deuterated methylarenes (compare entries 1 and 13). Such a result accords with the mechanism in Scheme 2 since only a small secondary kinetic isotope effect is expected for electron detachment. Indeed the absence of a deuterium isotope effect for electron transfer under these conditions coincides with the results of the photoionization measurements in Table I. Second, the use of substituted pyridine bases allowed the experimental rates to be varied over a wide range. Despite a ~3000-fold variation, the evaluated k_1 was independent of the structural effects of the pyridine bases, as shown in Table II, column 5.

3.1.2. The Rate Constant k_{-1} for Back Electron Transfer

The value of k_{-1} can be derived from the measured value of k_1. Since the ratio of rate constants represents the equilibrium constant for the electron exchange in Eq. 19, it is related to the overall free energy change and can be expressed as

$$\frac{k_1}{k_{-1}} = \exp[\mathcal{F}(E^0_{\mathrm{Fe}} - E^0_{\mathrm{Ar}})/RT] \tag{25}$$

where E^0_{Fe} and E^0_{Ar} are the standard oxidation potentials of FeL_3^{2+} and $ArCH_3$ in Eqs. 17 and 8, respectively, and $\mathcal{F}$ is the Faraday constant. This relationship provides the rate constant k_{-1} for back electron transfer from the values of k_1 in Table II together with the reversible potentials E^0_{Ar} and E^0_{Fe}.[27]

3.2. Mechanistic Formulation of Electron Transfer from Arenes

The oxidative substitution of arenes offer unique opportunity to examine the quantitative relationship between the rate and the driving force for electron transfer, since it is one of the few organic systems in

Table II. Electron-Transfer Rate Constants k_1 and k_{-1} (Back Electron Transfer) from Arenes to Iron(III) Oxidants.[a]

Methylarene	$(5X\text{-}phen)_3Fe^{3+}$ *X*	*Y-py* *Y*		k_1 $(M^{-1}s^{-1})$	k_{-1} $(M^{-1}s^{-1})$[b]
(HMB)	H	H	14(15)[c]	14.5^d	3.5×10^9
		4–CN	14		
		3–CN	15		
		2,6–Me	14		
	Cl	H		52	2.4×10^8
	NO_2	H		1300	1.2×10^8
(PMB)	H	H	0.28	0.28^d	5.1×10^9
		4–CN	0.23		
		3–Cl	0.30		
		2,6–Me	0.33		
	Cl	H		5.3	1.9×10^9
	NO_2	H		75	5.2×10^8
(DUR)	H	H	0.034(0.04)[c]	0.032^d	6.1×10^9
		4–CN	0.025		
		3–Cl	0.036		
		2,6–Me	0.026		
	Cl	H		3.8	1.4×10^{10}
	NO_2	H		16	1.2×10^9
(TMB)	H	H	0.0051	0.0048^d	1.4×10^9
		4–CN	0.0045		
		3–Cl	0.0041		
		2,6–Me	0.0055		
	Cl	H		2.8	1.6×10^{10}
	NO_2	H		4.1	4.4×10^8

[a]In acetonitrile containing 0.1 *M* tetraethylammonium or lithium perchlorate at 22°C with iron(III) oxidant.

[b]Determined from Eq. 25.

[c]Numbers in parentheses are the values of k_1 for the perdeuterated methylarene.

[d]Average value for the different pyridine bases.

which values of both the intrinsic rate constants [k_1, k_{-1}] and the free energy changes [$\mathcal{F}(E^0_{Ar} - E^0_{Fe})$] for electron transfer have been rigorously established. It is noteworthy that the rate constants k_{-1} for back electron transfer listed in Table II fall in the range closely approaching the diffusion-controlled limit of 10^9–10^{10} M^{-1} s^{-1}.[28] As such, neither k_{-1} nor k_1 can be considered to represent only a purely activation process for electron transfer; but they must also include diffusional processes. Let us therefore consider the general mechanism in which the measured rate constant k_1 for electron transfer takes into account both contributions, i.e.,

$$\frac{1}{k_1} = \frac{1}{k^\ddagger} + \frac{1}{k_r} + \frac{1}{k_p} \exp[\mathcal{F}(E^0_{Ar} - E^0_{Fe})/RT] \tag{26}$$

where $k^\ddagger$ represents the true activation rate constant for electron transfer, and k_r and k_p are the diffusion rate constants for the formation of the precursor complex and the dissociation of the successor complex, respectively.[29] Such a formulation derives from the overall electron transfer in Eq. 27 considered in terms of the three successive elementary steps outlined below.[30]

$$\text{overall:} \quad Ar + Fe(III) \underset{}{\overset{\Delta G_0}{\rightleftharpoons}} Ar^{\cdot+} + Fe(II) \tag{27}$$

$$\text{diffusion:} \quad \updownarrow w_r \qquad\qquad \updownarrow w_p$$

$$\text{activation:} \quad [Ar,Fe(III)] \overset{\Delta G_0'}{\rightleftharpoons} [Ar^{\cdot+},Fe(II)] \tag{28}$$

Scheme 3.

In Scheme 3 the diffusion rate constants k_r and k_p refer to the precursor and successor complexes (see brackets) in steady state, and the free energy change for the activation process in Eq. 28 is then given by $\Delta G_0'$ $= \Delta G_0 + w_p - w_r$. Since the arene is uncharged, we consider the reactant work term w_r to be nil. Accordingly the free energy change from the precursor to the successor complex in Eq. 28 is

$$\Delta G_0' = \mathcal{F}(E^0_{Ar} - E^0_{Fe}) + w_p \tag{29}$$

Table III. Free Energy Change, Activation Free Energy, and Intrinsic Barrier for Electron Transfer from Methylarenes to Iron(III) Oxidants.[a]

Methylarene	$(5X\text{-}phen)_3Fe^{3+}$ X	ΔE^0	$\Delta G_0^{'b}$	$\Delta G^{\ddagger c}$	$\Delta G_0^{\ddagger d}$	
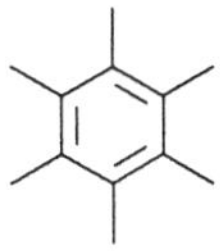 (HMB)	H	0.49	9.8	12.9	7.2	
	Cl	0.39	7.5	12.6	8.4	7.8
	NO_2	0.29	5.2	10.7	7.9	
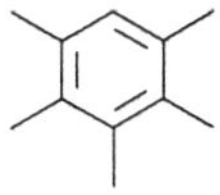 (PMB)	H	0.60	12.5	14.6	6.9	
	Cl	0.50	10.0	13.7	7.9	7.6
	NO_2	0.40	7.7	12.4	8.1	
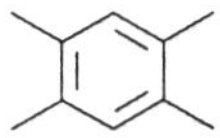 (DUR)	H	0.66	13.7	*e*	*e*	
	Cl	0.56	11.4	*e*	*e*	8.0
	NO_2	0.46	9.1	13.2	8.0	
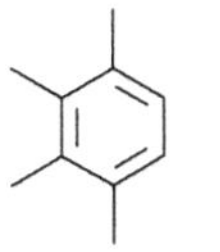 (TMB)	H	0.68	14.2	17.9	9.4	
	Cl	0.58	11.9	*e*	*e*	9.1
	NO_2	0.48	9.6	14.4	8.9	

[a]Energies in kcal mol^{-1}.
[b]Experimental uncertainty is ±0.5 kcal mol^{-1}; $\Delta G_0' = \mathcal{F}(E^0_{Ar} - E^0_{Fe}) + w_p$, with w_p taken as 1.5 kcal mol^{-1}.
[c]Experimental uncertainty ±0.2 kcal mol^{-1}.
[d]Evaluated from Marcus Eq. 31.
[e]Not evaluated because $k_1 > k_p \exp[\mathcal{F}(E^0_{Fe} - E^0_{Ar})/RT]$.

where w_p is the work term of the ion pair. The computed values of $\Delta G_0'$ are listed in Table III. The free energy of activation $\Delta G^\ddagger$ for electron transfer in Eq. 28 is evaluated from the rate constant by:

$$\Delta G^\ddagger = -RT \ln(k^\ddagger/Z) - w_r \tag{30}$$

where the collision frequency Z and the adiabaticity coefficient are taken to be 10^{11} M^{-1} s^{-1} and unity, respectively. The values of $\Delta G^\ddagger$ computed with the aid of Eqs. 26 and 30 are also listed in Table III.

3.3. Application of Marcus Theory to Arene Activation by Electron Transfer

The Marcus equation relates the activation free energy $\Delta G^{\ddagger}$ to the free energy change $\Delta G_0^{\prime}$ in the electron transfer step in Eq. 28 as

$$\Delta G^{\ddagger} = \Delta G_0^{\ddagger}\left[1 + \frac{\Delta G_0^{\prime}}{4\Delta G_0^{\ddagger}}\right]^2 + w_r \tag{31}$$

in which the intrinsic barrier $\Delta G_0^{\ddagger}$ represents the activation free energy for electron transfer when the driving force is zero, i.e., $\Delta G^{\ddagger} = \Delta G_0^{\ddagger}$ at $\Delta G_0^{\prime} = 0$. It is important to emphasize that the theoretical basis for the Marcus formulation of $\Delta G_0^{\ddagger}$ for Eq. 28 rests on an outer-sphere mechanism for electron transfer.

The last column in Table III lists the values of the intrinsic barriers for electron transfer computed from Eq. 31.[31] Since the electron exchange in the tris-phenanthrolineiron(III, II) redox couple is known to be very rapid,[32] the variations of $\Delta G_0^{\ddagger}$ in Table III largely reflect the changes associated with the conversion of the arene to its cation radical. Indeed for a given arene, the value of $\Delta G_0^{\ddagger}$ is found to be relatively invariant with the nature of the iron(III) oxidant. A close inspection of the values tabulated in Table III shows no consistent trend in $\Delta G_0^{\ddagger}$ of any significance. Moreover an average estimate of $\Delta G_0^{\ddagger}$ taking into account all the redox systems affords $\Delta G_0^{\ddagger} = 8.1$ kcal mol^{-1} with a standard deviation of only 0.7 kcal mol^{-1}, which is clearly within the experimental accuracy of the data.

The experimental results thus confirm the validity of Marcus theory (Eq. 31) to describe the free energy relationship for electron transfer. This conclusion is graphically illustrated in Figure 5, which presents a plot of $(\Delta G^{\ddagger})^{1/2}$ as a linear function of the driving force $(E^0_{Ar} - E^0_{Fe})$. Since the Marcus Eq. 31 can be rewritten in the form,

$$(\Delta G^{\ddagger})^{1/2} = (\Delta G_0^{\ddagger})^{1/2} + [w_p/4(\Delta G_0^{\ddagger})^{1/2}] + \mathcal{F}/4(\Delta G_0^{\ddagger})(E^0_{Ar} - E^0_{Fe}) \tag{32}$$

it indeed predicts the linear correlation in Figure 5.

3.3.1. Intrinsic Barrier to Electron Transfer from Arenes

The slope of the linear free energy relationship in Figure 5 is related to the intrinsic barrier $\Delta G^{\ddagger}$ (see Eq. 32). From the plot in Figure 5, a value

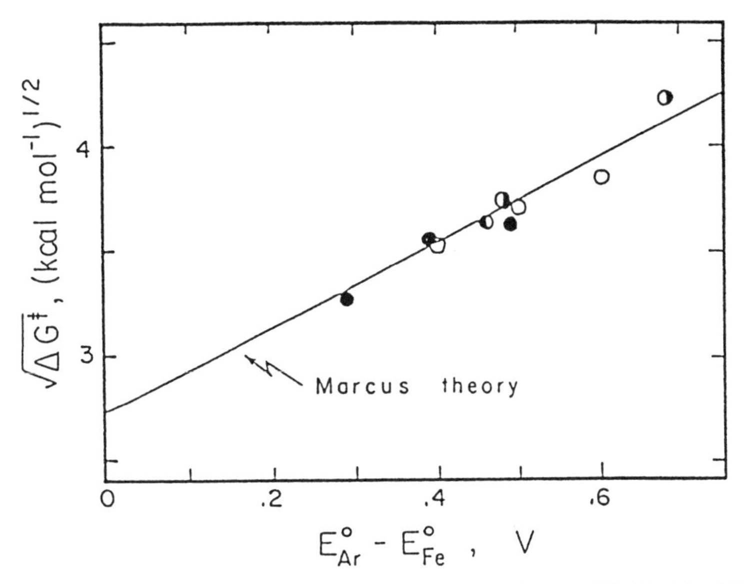

Figure 5. Fit of the Measured Rate Constants k_1 for HMB (•), PMB (o), DUR (◐), and TMB (◑) to the Marcus Equation (*full line*). The Line is the Fit to the Marcus Eq. 32 with $\Delta G_0^{\ddagger} = 8.2$ kcal mol^{-1}.

of $\Delta G^{\ddagger} = 8.2$ kcal mol^{-1} is obtained—in excellent agreement with those listed in the last column of Table III. Moreover the experimental value of the Brønsted α can be indicated as $\alpha \sim 0.7$ by taking the slope of the linear regression[33] of $\Delta G^{\ddagger}$ versus $\Delta G_0'$ in Table III. Such relatively large values of the Brønsted slopes (in the range of $\alpha \sim 0.8$) for electron transfer from arenes are clearly associated with driving forces in the endergonic region.[34] It is noteworthy that the Marcus relationship is quite effective in predicting the absolute magnitudes of the activation free energy as well as its variation with changes in the driving force for electron transfer. Central to the relationship is its ability to afford a reasonably consistent measure of the intrinsic barrier $\Delta G_0^{\ddagger}$ for electron transfer.

Because of its theoretical significance, we now wish to interpret the magnitude of the intrinsic barrier $\Delta G_0^{\ddagger}$ for electron transfer from arenes within the context of the Marcus rate theory. An important feature of Marcus theory is that it allows the prediction of the intrinsic barrier in terms of the reorganization energy, i.e., $\lambda = 4\Delta G_0^{\ddagger}$. For outer-sphere electron transfer, λ can be regarded simply as the sum of two contributions: $\lambda = \lambda_i + \lambda_o$.

The inner-sphere reorganization energy λ_i includes both reactants and takes into account the variations in their bond lengths, bond angles, and any specific interactions attendant on electron transfer. Since there are no appreciable differences in the experimental bond lengths and angles in tris-phenanthrolineiron(III, II),[32] λ_i can be regarded as negligible for the iron moiety. However, it is known that benzene undergoes a geometric change arising from the Jahn–Teller distortion in the cation radical that is generated on electron transfer.[35] An estimate of λ_i for the methylarenes examined in this study can thus be obtained from the value of the stabilization energy of $C_6H_6^{\cdot+}$ relative to the hypothetical cation radical in the nuclear configuration of benzene. Such a reorganization energy has been reported by Salem to be ~2.7 kcal mol^{-1}.[36]

The outer-sphere reorganization energy λ_o arises mainly from the changes in solvation (see Eq. 14), and it is evaluated from Marcus theory as

$$\lambda_o = 3.35 \times 10^2 \left[\frac{1}{2r_{Ar}} + \frac{1}{2r_{Fe}} - \frac{1}{d^\ddagger}\right]\left(\frac{1}{\eta^2} - \frac{1}{D}\right) \tag{33}$$

where r_{Ar} and r_{Fe} are the radii of the reactants considered as hard spheres, and $d^\ddagger \cong (r_{Ar} + r_{Fe})$. The index of refraction η and the static dielectric constant D of the solvent acetonitrile are 1.344 and 37.5, respectively. Taking r_{Fe} = 7 Å for tris-phenanthrolineiron(III) complexes[37] and r_{Ar} = 3.5 Å for the benzene derivatives, we calculate λ_o from Eq. 33 to be 21 kcal mol^{-1}. The results illustrated in Figure 3 and analyzed in Eq. 15 indicate that solvation energies are constant for the methylarene cation radicals examined in this study.

The total reorganization energy obtained as the sum of the inner-and outer-sphere contributions can thus be estimated as 24 kcal mol^{-1}. The resulting predicted value of the intrinsic barrier $\Delta G_0^\ddagger$ is 6 kcal mol^{-1}, which is in reasonable agreement with the measured value of 8.1 ± 0.7 evaluated from Eq. 31 (vide supra).

3.3.2. Diffusional Contribution to Endergonic Electron Transfer from Arenes

To relate the experimental free energy relationship for electron transfer to the theoretical relationship developed by Marcus, the diffusion rate constants must be explicitly included, as in Eq. 26. In particular, for endergonic processes of the magnitude encountered in this study, the diffusion rate constant k_p of the product pair can contribute as much as 25% to the measured rate constant k_1 for electron transfer. Therefore let us consider the general description of the rate constant k_1 when the driving force ΔG_0 is zero to the endergonic limits in terms of contributions from the activation rate constant $k^{\ddagger}$ and the diffusion rate constant k_d. Figure 6a shows the variation of $k^{\ddagger}$ with the driving force $\Delta G_0'$ according to Marcus theory (Eq. 31) for various values of $\Delta G_0^{\ddagger}$. Figure 6b represents the corresponding variations of the diffusion rate constant k_d obtained from Eq. 26 when $k^{\ddagger} >> k_r = k_p$, i.e.,

$$k_d^{-1} = k_r^{-1}\,[1 + \exp(\Delta G_0/RT)] \tag{34}$$

Figure 6c illustrates how the measured rate constant k_1 (where $1/k_1 = 1/k^{\ddagger} + 1/k_d$) varies with the driving force, for various values of $\Delta G_0^{\ddagger}$. To stress the importance of the diffusional contribution in our system, we have represented the k_1 relationship as the heavy line and the corresponding $k^{\ddagger}$ relationship as the dashed line. In each case we employed the average $\Delta G_0^{\ddagger} = 8.1$ kcal mol^{-1} from data as deduced from the Marcus equation. The diffusional contribution, represented by the gap between $k^{\ddagger}$ and k_1, is clearly seen to increase as one proceeds in our system into endergonic region beyond $\Delta G_0' \sim 10$ kcal mol^{-1}. It is therefore necessary to develop some operational criteria to determine when the diffusional contribution must be explicitly taken into account in the rate process for electron transfer.

An endergonic system (e.g., as applied in the Hammond postulate) is a thermodynamic concept and is of only limited utility in the context of the rate processes of importance here. Let us therefore refer to the endergonic *limits* of the kinetics, at which the rates of electron transfer are diffusion controlled. In this context, the endergonic *region* refers to driving forces at which the measured rate constant k_1 contains diffusion components (k_p and k_r) in addition to the activation component ($k^{\ddagger}$). The driving forces with $\Delta G_0' >> 0$ as the only criterion of endergonicity is inadequate, since the family of curves in Figure 6c clearly shows that the approach of k_1 to the diffusion limits is also strongly dependent on the magnitude of the intrinsic barrier $\Delta G_0^{\ddagger}$ for the particular system. For example, the endergonic limit is reached at $\Delta G_0' \cong 4$ kcal mol^{-1} when $\Delta G_0^{\ddagger} = 5$ kcal mol^{-1}, but the same limit is reached later at 16 kcal mol^{-1} when $\Delta G_0^{\ddagger}$ is doubled. In fact, Figure 7 illustrates how the magnitude of the intrinsic barrier determines the location of the endergonic region. [The curve represents systems with rate constants calculated from Eq.

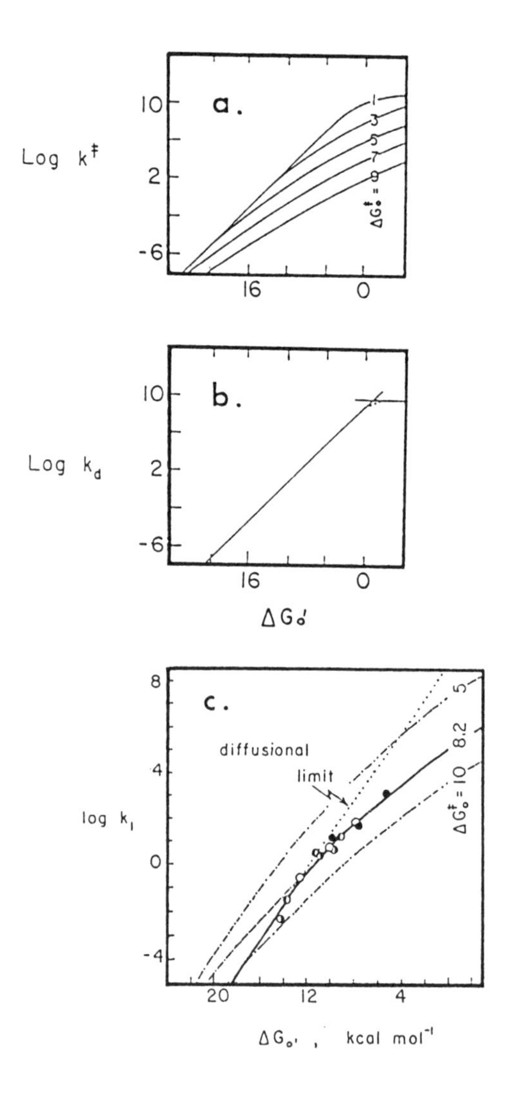

Figure 6. The Functional Forms of the Free Energy Relationships for the Rate Constants (a) $k^{\ddagger}$, (b) k_d, and (c) k_1. (*a*) Theoretical Curves Drawn According to Marcus Eq. 31 at Various Values of $\Delta G_0^{\ddagger}$. (*b*) Hypothetical Curve Drawn According to Eq. 34. (*c*) Curves Drawn with Various $\Delta G_0^{\ddagger}$ According to Eqs. 26 and 32. The Heavy Solid Curve Represents the Fit to the Experimental Data for HMB (•), PMB (○), DUR (◓), and TMB (◑) with $\Delta G_0^{\ddagger} = 8.2$ kcal mol^{-1}. Free Energy Relationships Drawn According to Marcus Eq. 31 Only, with $\Delta G_0^{\ddagger} = 5$, 8.2, or 10 kcal mol^{-1}.

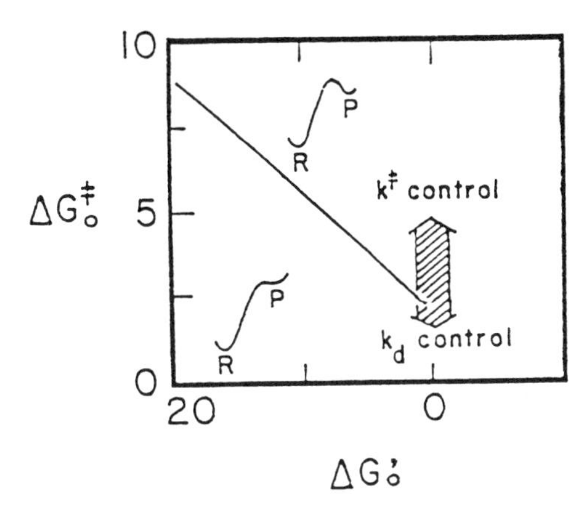

Figure 7. The Location of the Endergonic Region as a Function of the Intrinsic Barrier. The Curve is Arbitrarily Chosen to Consist of Equal Activation ($k^{\ddagger}$) and Diffusion (k_d) Components. R and P in the Reaction Diagrams Shown in the Inset Refer to the Reactants and Products, Respectively.

26 to consist of (arbitrarily) equal contributions from diffusion and activation. As the systems are displaced above this line, they are increasingly activation controlled, and those below the line are increasingly dominated by diffusion.] Since the curve in Figure 7 beyond $|\Delta G_0^{\circ}| > 1$ kcal mol^{-1} is reasonably straight, we can conclude that the location of the endergonic region varies more or less linearly with the magnitude of $(\Delta G_0^{\circ}/\Delta G_0^{\ddagger})$, i.e., the driving force normalized to the intrinsic barrier. Unfortunately such a criterion is not easy to apply in practice.

Accordingly, let us consider an alternative description of the endergonic limits in terms of rate-limiting diffusional processes k_p and k_r with a Brønsted slope of 1 (see Figure 6b). The endergonic region is then defined as that in which the Brønsted slope is ~1.[34] To apply Brønsted slopes as a quantitative criterion, the dependence of the experimental rate constant k_1 (which includes diffusion) on the driving force, i.e., $\partial \ln k/\partial \Delta G_0^{\circ}$, must be evaluated separately from that involving the activation (theoretical) rate constant $k^{\ddagger}$. The corresponding values of the experimental and theoretical Brønsted slopes α_1 and $\alpha^{\ddagger}$, respectively, are plotted as a function of the driving force for arene oxidations in Figure 8. The theoretical Brønsted slopes $\alpha^{\ddagger}$ were computed from the Marcus Eq. 31 since it provides a reasonable fit to the experimental data (vide supra). The junction at which the α_1 and $\alpha^{\ddagger}$ curves diverge shows that

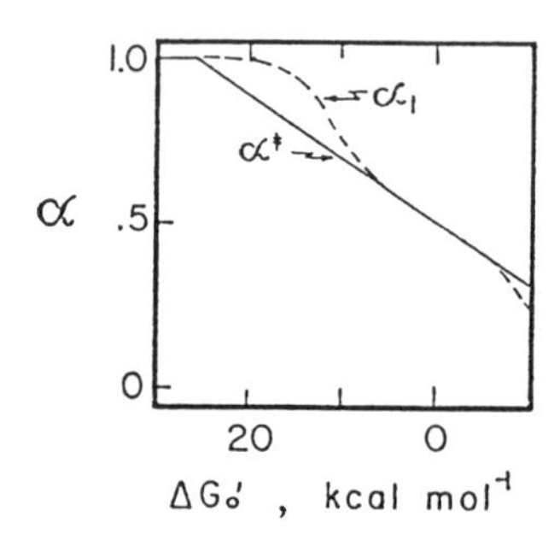

Figure 8. The Discrepancy Between the Measured (α_1) and the Theoretical ($\alpha^{\ddagger}$) Brønsted Slopes as a Function of the Driving Force for Electron Transfer. The Values of α_1 Were Derived from Eq. 26, and $\alpha^{\ddagger}$ Were Computed from Marcus Eq. 32 with $\Delta G_0^{\ddagger} = 8.2$ kcal mol^{-1}.

complete activation control of the electron transfer rates is observed only within a rather narrow window of the driving force, $\Delta G_0' \cong \pm 5$ kcal mol^{-1}, in the endergonic region. The widening gap beyond this point represents increasing control of the electron transfer rate by diffusional processes. Since the marked discrepancies between the experimental α_1 and the theoretical $\alpha^{\ddagger}$ characterize the endergonic region, we propose the magnitude of the Brønsted slope to be a useful indicator for rate processes in which diffusion must be explicitly taken into account.

At this juncture, however, we must clearly emphasize that any theoretical argument that relates the magnitude of the Brønsted slope to transition state structures or diagrams (such as those represented in Figure 7) refers specifically to values of $\alpha^{\ddagger}$ and not α_1. Since only α_1 is accessible by experiment, it is important to take explicit cognizance of diffusional contributions when rationalizing the experimental results.[38] This caveat is particularly pertinent to organic systems in which rates of electron transfer are likely to lie in the endergonic region because of limiting values of the standard redox potentials and reorganization energies.

4. SIDE CHAIN SUBSTITUTION OF METHYLARENES BY ELECTRON TRANSFER: THE DEPROTONATION OF CATION RADICAL INTERMEDIATES

Methylarenes $ArCH_3$ are known to undergo oxidative substitution reactions via the arene cation radical formed by either chemical or electrochemical methods.[39] In particular, these arenes are subject to oxidative degradation of the methyl side chain, as in the industrially important cobalt-catalyzed conversion of *p*-xylene to terephthalic acid.[40] The results of numerous chemical and electrochemical studies are compatible with the initial steps, which can be outlined in general form as

$$ArCH_3 \rightleftharpoons ArCH_3^{\cdot+} + [e] \quad (35)$$

$$ArCH_3^{\cdot+} \rightleftharpoons ArCH_2^{\cdot} + H^+ \quad (36)$$

$$ArCH_2^{\cdot} \rightarrow [e] + ArCH_2^{+}\text{, etc.} \quad (37)$$

Scheme 4.

where [*e*] represents the redox couple, i.e., either the electrode or the reduced oxidant such as Fe(III) → Fe(II) and Co(III) → Co(II). Scheme 4 is a classical example of an ECE process commonly encountered in the electrochemical literature.[41] As such, the degradation of the methyl side chain commences by prior electron transfer in Eq. 35, followed by the critical loss of an α-proton from the radical cation $ArCH_3^{\cdot+}$ in Eq. 36.

4.1. Rates of Deprotonation of Methylarene Cation Radicals

The oxidative substitution of methylarenes based on the iron(III) oxidants in Scheme 2 also presents the opportunity to examine quantitatively the dynamics of proton transfer from the methylarene cation radical. Thus the rate constant k_e for iron(III) reduction in Eq. 24 includes the second-order rate constant k_2 for the deprotonation of the transient methylarene cation radical by various pyridine bases, i.e.,

$$ArCH_3^{\cdot+} + \text{Y–py} \xrightarrow{k_2} ArCH_2^{\cdot} + \text{Y–pyH}^+ \quad (38)$$

and it can be extracted from the rate data of the type illustrated in Figure 4.[42] Thus it follows from Eq. 24 that the slope represents the ratio of the rate constants k_{-1}/k_1k_2. This together with the relationship in Eq. 25 yields the value of k_2 for the deprotonation of the cation radical $ArCH_3^{\cdot+}$. For the more reactive methylarenes–pyridine systems, we also developed a slightly modified, alternative procedure for the kinetics analysis, which

Table IV. Kinetic Acidity of Methylarene Cation Radicals with Different Substituted Pyridine Bases.[a]

Y-Pyridine		*Deprotonation Rate* ($\log k_2$)[b]					
Y	pK_a^B	$HMB^{\bullet+}$	d_{18}-$HMB^{\bullet+}$	$PMB^{\bullet+}$	$DUR^{\bullet+}$	d_{14}-$DUR^{\bullet+}$	$TMB^{\bullet+}$
2-Fluoro	4.2	3.95					
2-Chloro	6.3	4.80					
3-Cyano	7.0	5.00					
4-Cyano	8.0	5.28	4.63	5.90	6.05	—	5.51
3-Chloro	9.0	5.72	5.05	6.19	6.21	5.58	5.93
3-Fluoro	9.4	5.75					
Hydrogen	12.3	6.71	6.20	7.15	7.43	6.94	6.67
2-Methyl	14.0	7.14					
4-Methyl	14.3	7.01					
4-Methoxy	15.0	6.97					
2,6-Dimethyl	15.4	7.03 (7.32)[c]	6.46 (6.87)[c]	7.56 (7.60)[c]	8.05 (7.76)[c]	—	7.46
2,6-Di-*t*-butyl	11.8	4.94					
2,4,6-Trimethyl	16.8	7.39 (7.54)[c]					

[a]In acetonitrile containing 0.1 *M* tetraethylammonium perchlorate or lithium perchlorate, at 22°C.
[b]k_2 in $M^{-1}s^{-1}$.
[c]Values in parentheses from ref. 44.

is applicable over a wider range of pyridine concentrations. The analysis by multiple linear regression leads directly to the values of the electron transfer rate constant k_1 and to the ratio of rate constants k_{-1}/k_2. As expected, the values of k_1 are the same as those obtained from Eq. 24 and are included among those in Table II. Since the quotient of k_{-1}/k_2 and k_1 is equivalent to the slope in Figure 4, it also provides for the evaluation of the proton transfer rate constant k_2 with the aid of Eq. 25. The values of k_2 for the deprotonation of hexamethylbenzene cation radical by various pyridine bases are listed in Table IV. It is noteworthy that these deprotonation rate constants range from ~4 x 10^2 M^{-1} s^{-1} for the weakest base 2-fluoropyridine to more than 9 x 10^5 M^{-1} s^{-1} for the strongest base 2,4,6-trimethylpyridine examined in this system. The same procedure was used to determine values of k_2 for the other methylarene cation radicals, which are also included in Table IV.

4.2. Direct Observation of the Kinetic Acidities of Arene Cation Radicals

The second-order rate constants k_2, as evaluated in the preceding section, are critically dependent on the redox equilibrium that is set by the potential difference ($E^0_{Ar} - E^0_{Fe}$) between the arene donor and iron(III) oxidant according to Eqs. 24 and 25. Since this indirect method for the evaluation of k_2 is highly dependent on the accurate determination of the driving force,[43] we sought an alternative method based on the charge-transfer excitation of the electron donor–acceptor complex of the methylarenes with tetranitromethane to be described in more detail in Section 5. Suffice it to mention here that the photoexcitation ($h\nu_{CT}$) generated the methylarene cation radical in sufficient concentration for its direct observation by time-resolved spectroscopy, i.e.,[44]

$$\mathrm{ArMe} + \mathrm{C(NO_2)_4} \rightleftharpoons [\mathrm{ArMe}, \mathrm{C(NO_2)_4}]$$

$$\xrightarrow{h\nu_{CT}} [\mathrm{ArMe}^{\cdot +}, \mathrm{C(NO_2)_3^-}, \mathrm{NO_2}] \qquad (39)$$

where the brackets enclose species initially trapped in the solvent cage. Thus the cation radicals from HMB, PMB, and DUR showed characteristic absorption bands at λ_{max} 495, 485, and 440 nm, respectively, upon the CT excitation of the tetranitromethane complex. The spectral examination of [$HMB^{\cdot +}$] at the monitoring wavelength of 500 nm established the decay to follow second-order kinetics in dichloromethane on the return of the absorbance to the baseline. This kinetics behavior indicated that the first-formed ion radical pair in Eq. 39 was sufficiently long-lived to suffer diffusive separation (Eq. 40) prior to the second-order recombination (Eq. 41).

$$[\mathrm{ArCH_3^{\cdot +}}, \mathrm{C(NO_2)_3^-}] \rightarrow \mathrm{ArCH_3^{\cdot +}} + \mathrm{C(NO_2)_3^-} \qquad (40)$$

$$\mathrm{ArCH_3^{\cdot +}} + \mathrm{C(NO_2)_3^-} \rightarrow \mathrm{ArCH_2^{\cdot}} + \mathrm{HC(NO_2)_3} \qquad (41)$$

The latter allowed the competition for $ArCH_3^{\cdot +}$ to occur with added pyridines in the concentration range of 5×10^{-4} to 5×10^{-2} M. Figure 9A typically illustrates the complete return of the $HMB^{\cdot +}$ absorbance to the baseline, with the excellent fit of the smooth computed curve to the experimental decay with first-order kinetics. The linear dependence of the observed first-order rate constant on the concentration of the various

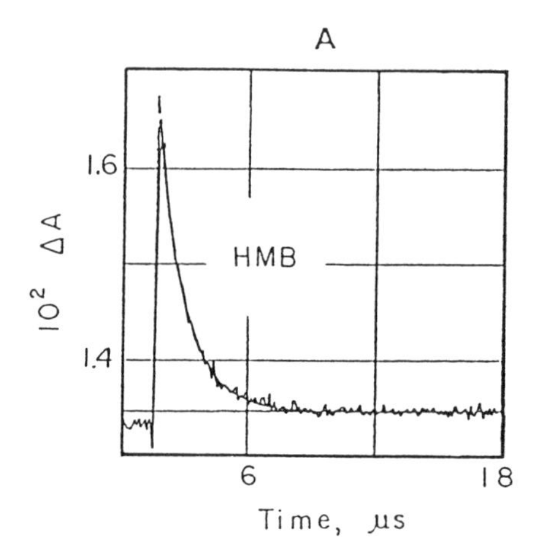

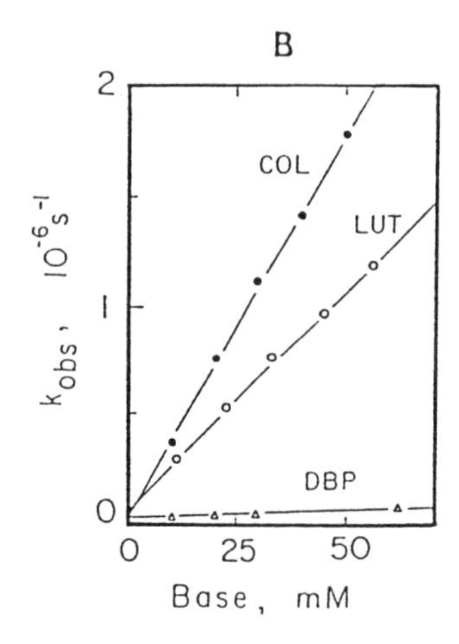

Figure 9. First-order Decay of the $HMB^{\cdot+}$ Absorbance at 495 mn Following the CT Excitation of 0.05 *M* HMB and 0.1 *M* TNM in Acetonitrile Containing 0.015 *M* Lutidine and 0.1 *M* $LiClO_4$ at 25°C. (*A*) Quenching of $HMB^{\cdot+}$ by Pyridine Bases, Showing the Dependence of the Pseudo-first-order Rate Constant k_{obs} from the 532-nm Excitation of 0.05 *M* HMB and 0.1 *M* TNM in Acetonitrile Containing 0.1 *M* $LiClO_4$ and COL (•), LUT (o), and DBP (Δ) (*B*).

pyridine bases in Figure 9B yielded the second-order rate constant k_H for the quenching of $ArCH_3^{\cdot+}$ by deprotonation, i.e.,

$$\mathrm{C_6(CH_3)_6^{\cdot+}} + \text{2,6-lutidine} \xrightarrow{k_2} \mathrm{C_6(CH_3)_5CH_2^{\cdot}} + \text{2,6-lutidine-H}^{+} \qquad (42)$$

The deuterium kinetic isotope effect of $k_H/k_D = 2.8$ obtained from the comparison of k_H for HMB with that of HMB-d_{18} confirmed the proton transfer in Eq. 42 and similar pseudo-first-order kinetics behavior was observed with the other methylarenes. Most importantly, the values of the second-order rate constants k_2 listed in Table IV (see the values in parentheses) are in agreement with those obtained by the indirect method. Thus the values of the kinetic acidities ($\log k_2$) obtained from the iron(III) oxidation of methylarenes are within the experimental uncertainty the same as those measured directly from the observation of the methylarene cation by time-resolved spectroscopy.

4.3. Free Energy Relationship for the Deprotonation of Methylarene Cation Radicals

The extensive set of kinetic data in Table IV for the various methylarene cation radicals with different substituted pyridine bases offers the opportunity to examine the quantitative relationship between the rate and the driving force for proton transfer from a carbon acid. Thus for a particular methylarene cation radical, the activation free energy for proton transfer to various pyridine bases as given by $\Delta G^{\ddagger} = -RT \ln(k_2/Z)$ in Table V follows the monotonic trend shown by the family of curves in Figure 10.[43] As in Scheme 3 the corrected driving force $\Delta G_0' = \Delta G_0 - w_r + w_p$ with $\Delta G_0 = RT \ln 10(pK_a^A - pK_a^B)$. The acidity constants pK_a^A and pK_a^B refer to the methylarene cation radical $ArCH_3^{\bullet+}$ and the pyridine conjugate acid pyH^+, respectively.[45] To collect the work terms, a corrected acidity constant can be defined as $pK_{a'}^A = pK_a^A + (w_p - w_r)/RT \ln 10$, so that $\Delta G_0' = RT \ln 10(pK_{a'}^A - pK_a^B)$.

Although the statement of Marcus theory in Eq. 31 is theoretically based on outer-sphere electron transfer, this functional form has been used to predict the rates of such inner-sphere processes as proton and methyl transfer as well as S_N2 substitutions.[46,47] Accordingly let us examine how far the Marcus equation can be extended to accommodate the proton transfers from methylarene cation radicals.[48] Since the ap-

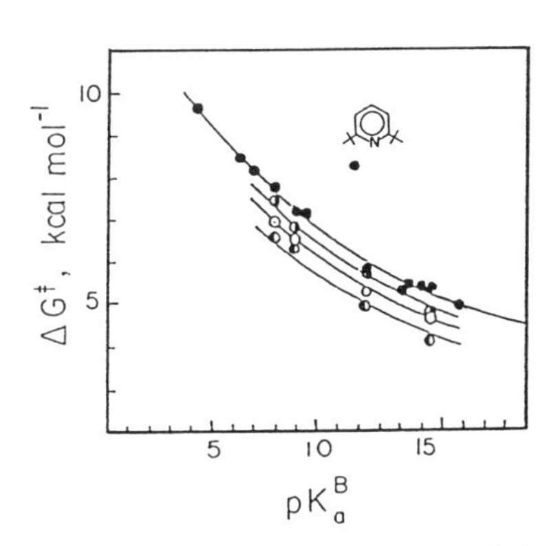

Figure 10. Relationship of the Free Energy of Activation for Proton Transfer to the pK_a^B of the Pyridine Base for HMB (•), PMB (○), DUR (◐), and TMB (◑).

Table V. Activation Free Energies for the Deprotonation of Methylarene Cation Radicals with Pyridine Bases.[a]

Y-pyridine		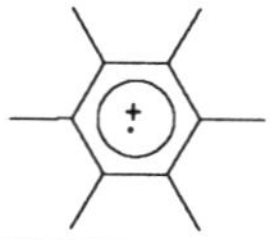		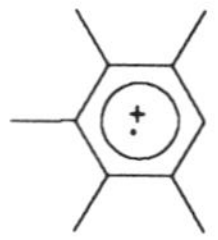		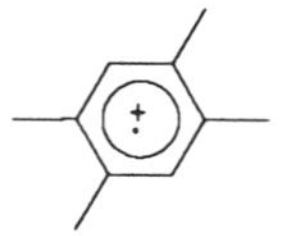			
Y	pK_a^B	$\Delta G^{\ddagger}$	$\Delta G^{\ddagger}/\Delta G_0^{\ddagger}$	$\Delta G^{\ddagger}$	$\Delta G^{\ddagger}/\Delta G_0^{\ddagger}$	$\Delta G^{\ddagger}$	$\Delta G^{\ddagger}/\Delta G_0^{\ddagger}$	$\Delta G^{\ddagger}$	$\Delta G^{\ddagger}/\Delta G_0^{\ddagger}$
2-F	4.2	9.58	1.50						
2-Cl	6.3	8.42	1.32						
3-CN	7.0	8.15	1.27						
4-CN	8.0	7.77	1.21	6.93	1.08	6.72	1.05	7.46	1.17
3-Cl	9.0	7.17	1.12	6.53	1.02	6.51	1.02	6.89	1.08
3-F	9.4	7.13	1.11						
H	12.3	5.83	0.91	5.23	0.82	4.85	0.76	5.88	0.92
2-Me	14.0	5.24	0.82						
4-Me	14.3	5.42	0.85						
4-MeO	15.0	5.47	0.86						
2,6-di-Me	15.4	5.39	0.84	4.67	0.73	4.01	0.63	4.81	0.75
2,6-di-*t*-Bu	11.8	8.23	1.29						
2,4,6-tri-Me	16.8	4.90	0.77						

[a]In kcal mol^{-1}, from k_2 values in Table IV. $\Delta G_0^{\ddagger}$ = 6.4 kcal mol^{-1} for all the methylarenes.

plication of Eq. 31 to proton transfer requires a knowledge of three unknowns, namely, $pK_{a'}^{A}$, w_r, and $\Delta G^{\ddagger}$, we begin by reformulating it as

$$\Delta G^{\ddagger} = A + B(pK_a^B) + C(pK_a^B)^2 \qquad (43)$$

where $A = w_r + \Delta G_0^{\ddagger} + RT\ln 10[1 + RT\ln 10(pK_{a'}^{A})/\ 8\,\Delta G_0^{\ddagger})](pK_{a'}^{A})/2)$,B= $-RT\ln 10[\frac{1}{2} + RT\ln 10 - (pK_{a'}^{A}/8\Delta G_0^{\ddagger})]$, and $C = (RT\ln 10)^2/\ (16\Delta G_0^{\ddagger})$. Considered in this way, the curvatures of the plots in Figure 10 are dependent only on the value of the intrinsic barrier $\Delta G_0^{\ddagger}$ for proton transfer. The treatment of the data in Table IV for hexamethylbenzene cation radical by a quadratic regression analysis affords $\Delta G_0^{\ddagger} = 6.4$ kcal mol^{-1} (not including the datum for 2,6-di-*t*-butylpyridine). Indeed the similarity of the plots in Figure 10 to the curvature for $HMB^{\cdot +}$ suggests that $\Delta G_0^{\ddagger}$ is rather constant for the series of methylarene cation radicals in Table V.

The Marcus relationship also leads to an evaluation of the acidity constants $pK_{a'}^{A}$ for the various methylarene cation radicals by the introduction of the value for the intrinsic barrier $\Delta G_0^{\ddagger}$ in the coefficient *B*, as obtained from the quadratic regression analysis of Eq. 43. Moreover the introduction of the values of both $\Delta G_0^{\ddagger}$ and $pK_{a'}^{A}$ in the coefficient *A* leads to a value for the work term w_r = 4.6 kcal mol^{-1} for HMB. These are collected in Table VI for the series of methylarene cation radicals examined in this study. Figure 11 shows the fit of the experimental data in Table IV (ordinate) with the parameters in Table VI (abscissa) as they are related by the Marcus equation (normalized to the intrinsic barrier).[51] It is interesting to note the trend in Table VI for the value of the work term w_r to decrease with the number of methyl substituents in $ArCH_3$. Furthermore the magnitude of the variation in the work terms, i.e. $(w_p - w_r)/RT\ln 10$, largely overwhelms the variation in the acidity constant pK_a^A. As a result, the corrected acidity constant $pK_{a'}^{A}$ actually follows a trend opposite to that of acidity constant pK_a^A (vide infra).

The plot in Figure 11 shows a Brønsted slope varying between α = 0.30–0.15 for the proton transfer from various methylarene cation radicals to the different pyridine bases. Slopes of such magnitudes correspond to an overall free energy change lying in the exergonic region.[34] This is in agreement with the values of $pK_{a'}^{A}$ in Table VI. Indeed the kinetics result of the deuterium isotope effect in Table VII leads to the same conclusion. Thus Figure 12 shows the deuterium kinetic isotope effect for proton transfer to decrease with an increasing driving force.

Table VI. Marcus Parameters for the Deprotonation of Methylarene Cation Radicals by Pyridine Bases.

$ArCH_3^{\bullet +}$	w_r (kcal mol^{-1})	$pK_a^{A'}$	w_p[a] (kcal mol^{-1})
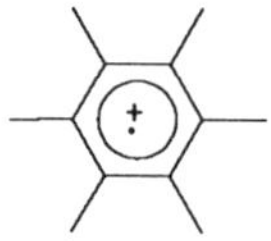	4.6	2.0	4.0
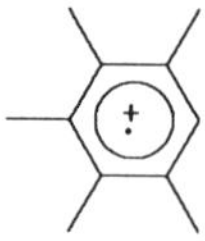	3.9	2.0	9.3
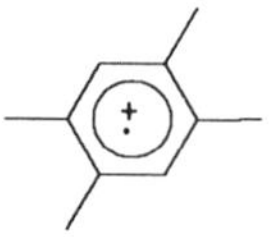	2.7	3.8	12
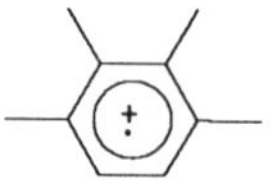	4.0	3.0	13

[a]Evaluated from the relationship $w_p = w_r + RT\ln 10(pK_a^{A'} - pK_a^{A})$, where the pK_a^{A} values are derived from thermochemical calculations (see text).

Table VII. Deuterium Kinetic Isotope Effects $k_2(H)/k_2(D)$ for the Deprotonation of Methylarene Cation Radicals by Different Substituted Pyridine Bases.[a]

 Y	pK_a^B	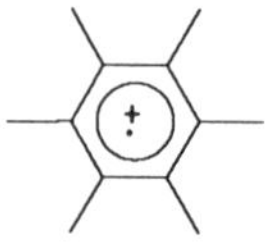 $k_2(H)/k_2(D)$	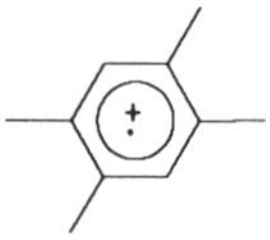 $k_2(H)/k_2(D)$
4-Cyano	8.0	4.4	—
3-Chloro	9.0	4.7	3.5
Hydrogen	12.3	3.6	2.6
2,6-Dimethyl	15.4	3.9	—

[a]From the data in Table IV.

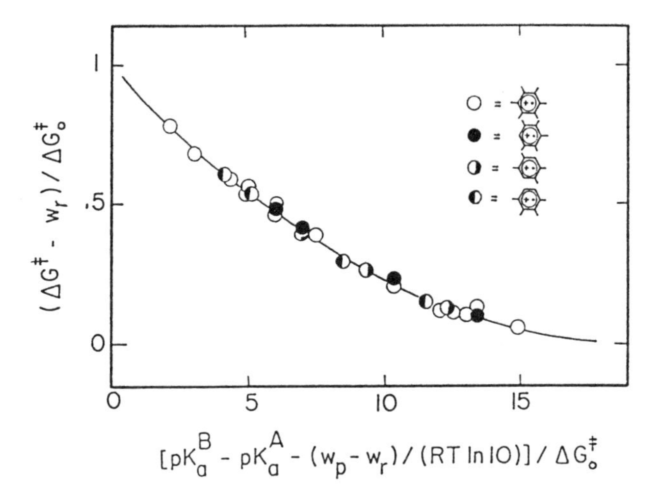

Figure 11. Unified Correlation of the Activation Free Energy and the Corrected Free Energy Change for the Deprotonation of the Cation Radicals for HMB (○), PMB (•), DUR (◑), and TMB (◐). The Theoretical Variation Calculated by the Marcus Equation is Indicated by the Line.

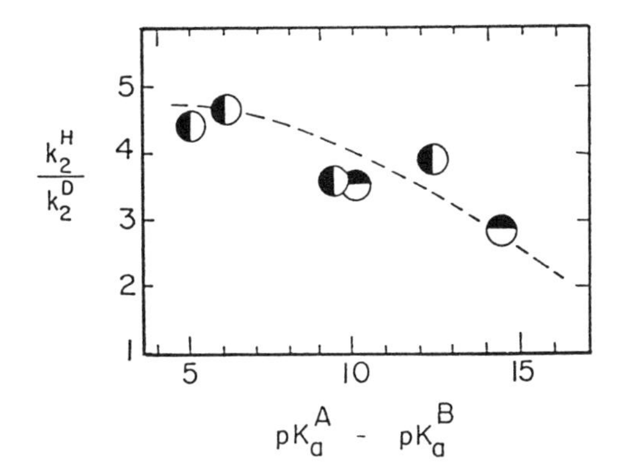

Figure 12. Deuterium Kinetic Isotope Effect as a Function of the Driving Force for Proton Transfer from the Cation Radicals of HMB (◐) and DUR (⊖).

We interpret the magnitude of the Brønsted slope and the trend in the kinetic isotope effect to reflect an early transition state in which proton transfer has not proceeded beyond the symmetrical situation. Such a qualitative description of methylarenes as acids accords with the relatively low sensitivity of proton transfer rates in Figure 11 to the steric effects of the pyridine bases. Thus a pair of *ortho* or 2,6-di-*t*-butyl substituents as in 2,6-di-*t*-butylpyridine is required to significantly perturb the Brønsted correlation of various pyridine bases (see Figure 10). Furthermore the analysis from the Marcus approach reveals the high degree to which w_r and w_p contribute to the rates of proton transfer in comparison with those relevant to (outer-sphere) electron transfer in which their neglect does not materially affect the correlation (see Figure 5). We hasten to add the caveat that the theoretical basis for the inner-sphere work terms w_r and w_p in proton transfer is complex.[48] Indeed the proton transfer between methylarene cation radicals and pyridine bases represents a rather unusual situation insofar as acid–base reactions are concerned. Thus the reactant pair is strikingly akin to the product pair, especially if viewed in terms of the following charge-transfer interactions:

$$[A^{+\cdot}\ D] \overset{\Delta G^{\circ\prime}}{\rightleftharpoons} [D\cdot\ A^{+}] \quad (44)$$

Thus proton transfer within the encounter complex is accompanied by an interchange of the π-donor (D) and π-acceptor (A) capacities of the methylarene and pyridine moieties on conversion to the successor complex in Eq. 44.[53,54] As such, it is possible that the magnitudes of the work terms w_r and w_p in <u>this</u> system are comparable.

As useful as the Marcus equation is in correlating all the rates of proton transfer in Figure 11 it does not yield directly an independent value for the acidity constant pK_a^A. To evaluate the acidity of methylarene cation radicals, we rely on the thermochemical calculations by Nicholas and Arnold,[55] which lead to the general relationship $pK_a^A = -(\mathcal{F}E^0_{Ar} + \Delta G_0^{CH}) / RT \ln 10 + \text{constant}$, where ΔG_0^{CH} is the relevant C–H bond energy in the methylarene. In the series of cation radicals examined in this study, the variation in ΔG_0^{CH} is minor, and we take $pK_a^A = -(\mathcal{F}/RT \ln 10)E^0_{Ar}$ + constant. An estimate of $pK_a^A = 0$ for $HMB^{\cdot+}$ is obtained from this relationship and the data in Table I, if the acidity constant is taken as –12 for the toluene cation radical, as evaluated by Arnold and Nicholas.[56] Interestingly, the value compares with $pK_a^A = 2$ for $HMB^{\cdot+}$ from the Marcus relationship expressed as Eq. 43.

5. SIDE CHAIN VERSUS NUCLEAR SUBSTITUTION OF METHYLARENE CATION RADICALS

Two major pathways have been identified in the oxidation of arenes, namely side chain and nuclear substitution as given below with methoxytoluene as an illustrative example.[57] This dichotomy has been identified in most of the previous studies of aromatic oxidations by extensive product studies using various types of alkylaromatic hydrocarbons.[58]

$$MeO-C_6H_4-CH_3 \xrightarrow[HOAc]{[-2e]} MeO-C_6H_4-CH_2OAc + 2H^+$$

(side chain substitution)

$$MeO-C_6H_4-CH_3 \xrightarrow[HOAc]{[-2e]} MeO-C_6H_3(OAc)-CH_3 + 2H^+$$

(nuclear substitution)

The oxidation of *p*-methoxytoluene (PMT) by tris-phenanthroline-iron(III) in acetonitrile solution containing pyridine as an added base is accompanied by the rapid change in color from blue to red. The distinctive color change is indicative of the reduction of the iron(III) complex $Fe(phen)_3^{3+}$ to the iron(III) complex $Fe(phen)_3^{2+}$.[59] The stoichiometry for the iron(III) oxidation of *p*-methoxytoluene corresponds to an overall 2-electron oxidation, viz.,

$$p\text{-}CH_3C_6H_4OCH_3 + 2Fe(phen)_3^{3+} \xrightarrow{[B]} [p\text{-}CH_3C_6H_4OCH_3]_{ox} + 2Fe(phen)_3^{2+} \quad (45)$$

where the quantity in brackets represents the oxidation products of *p*-methoxytoluene illustrated below:

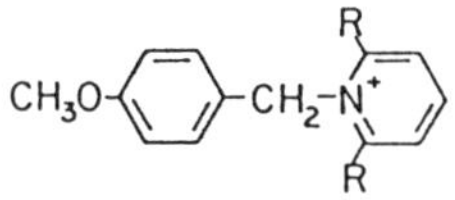

Ia, R = H **IIa**, R = H

Ib, R = CH_3 **IIb**, R = CH_3

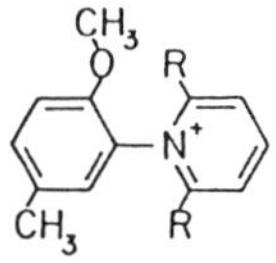

The base-promoted oxidation of *p*-methoxytoluene by iron(III) is unusual for two principal reasons. First, the results in Table VIII indicate that the products of oxidation that derive from either side chain and nuclear substitution (as exemplified by structures **I** and **II**, respectively) are highly dependent on whether pyridine or 2,6-lutidine is employed. Second, the results in Table IX indicate that the kinetic isotope effect measured with *p*-methoxytoluene-α,α,α-d_3 is either relatively large or small depending on whether 2,6-lutidine or pyridine is the base. Since both problems relate to the nature of the base B, its role in the oxidative rate process must be defined quantitatively. Indeed, the observed products, stoichiometry, kinetics, and kinetic isotope effects are akin to those derived from earlier studies of *p*-methoxytoluene oxidation,[60] in which the cation-radical [$PMT^{\bullet +}$] is the reactive intermediate formed by one-electron transfer. The validity of this premise is demonstrated by the

Table VIII. Oxidation Products of *p*-Methoxytoluene by tris-Phenanthrolineiron(III) in the Presence of Pyridine and 2,6-Lutidine as Added Bases.[a]

	p-Methoxytoluene		$Fe(phen)_3^{3+}$		
Base [B] (mmol)	*Initial (mmol)*	*Final (mmol)*	*Initial (mmol)*	*Final (mmol)*	*Products*[b] *mol (%)*
Pyridine (1.01)	0.50	0.11	1.00	0.05	**Ia** 0.045(9) **IIa** 0.35 (69)
2,6-Lutidine (1.00)	0.50	0.25	1.00	0.33	**Ib** 0 (<3) **IIb** 0.19 (38)

[a]In 20 ml of acetonitrile at 25°C with $Fe(phen)_3(PF_6)_3$.

[b]Yield based on initial PMT. Estimated error of ±10% in the NMR analysis.

Table IX. Comparative Rates of Oxidation of *p*-Methoxytoluene and its α,α,α-Trideuterio Derivative with $Fe(phen)_3^{3+}$.[a]

Base *[B]*	k_H $(M^{-1}s^{-1})$	k_D $(M^{-1}s^{-1})$
Pyridine	$1.7 \pm 0.2 \times 10^{-3}$	$6.2 \pm 0.2 \times 10^{-4}$
2,6-Lutidine	$3.7 \pm 0.2 \times 10^{-3}$	$6.7 \pm 0.2 \times 10^{-4}$

[a]In acetonitrile solutions containing ~10^{-2} *M* base, 10^{-2} *M* *p*-methoxytoluene, 10^{-4} *M* $Fe(phen)_3(PF_6)_3$, and 0.1 *M* $LiClO_4$ at 22°C. The rate constants k_H and k_D refer to $k_e[Fe(III)]_0/[PMT]_0[B]_0$ for *p*-$CH_3C_6H_4OCH_3$ and *p*-$CD_3C_6H_4OCH_3$, respectively.

observation of the well-resolved esr spectrum of $[PMT^{\cdot+}]$ shown in Figure 13 simply on the exposure of *p*-methoxytoluene to $Fe(phen)_3^{3+}$. The basic outline of such an electron transfer mechanism as applied to the iron(III) oxidations pertinent to this study is presented below, and each step is labeled with its appropriate rate constant.

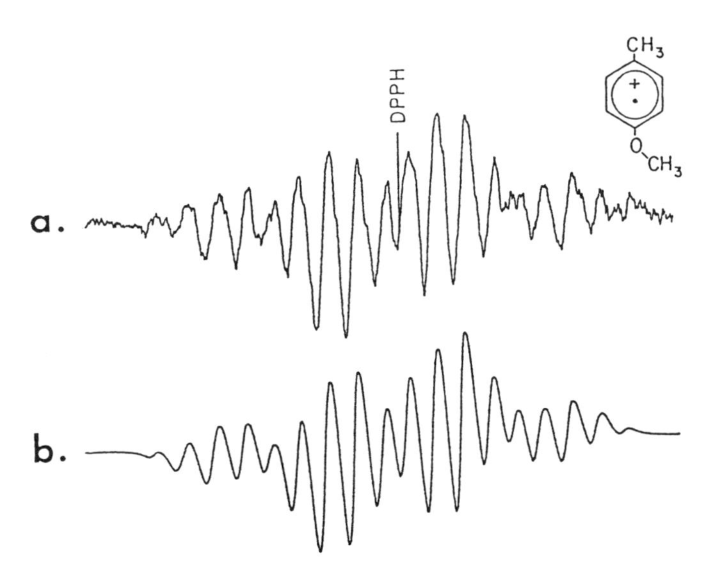

Figure 13. The X-band esr Spectrum of $[PMT^{\cdot+}]$ Obtained from the Thermal Reaction of *p*-Methoxytoluene and $Fe(phen)_3^{3+}$ (*top*). Computer simulated spectrum of $[PMT^{\cdot+}]$ (*bottom*).

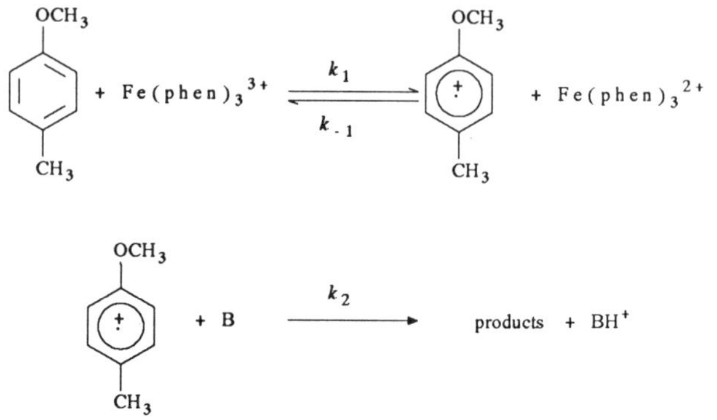

$$p\text{-}CH_3OC_6H_4CH_3 + Fe(phen)_3^{3+} \underset{k_{-1}}{\overset{k_1}{\rightleftharpoons}} p\text{-}CH_3OC_6H_4CH_3^{\cdot+} + Fe(phen)_3^{2+} \quad (46)$$

$$p\text{-}CH_3OC_6H_4CH_3^{\cdot+} + B \xrightarrow{k_2} \text{products} + BH^+ \quad (47)$$

Scheme 5.

The role of the base B in Scheme 5 is to displace the redox equilibrium in Eq. 46 by proton removal in the followup reaction in Eq. 47. Using the equilibrium state approximation for [$PMT^{\cdot+}$] as the transient intermediate in Scheme 5 leads to the following rate law (compare Eq. 22):

$$-\ln\frac{[Fe(III)]}{[Fe(III)]_0} - 1 + \frac{[Fe(III)]}{[Fe(III)]_0} = 2\,\frac{k_1k_2}{k_{-1}}\,\frac{[PMT]_0[py]_0}{[Fe(III)]_0}\,t \quad (48)$$

where the brackets represent concentrations at time t, and the subscript 0 refers to the initial condition at $t = 0$. Comparison of the rate law in Eq. 48 with the experimental kinetics leads to the relationship among the various rate constants as:

$$k_e = \frac{2k_1k_2}{k_{-1}}\,\frac{[PMT]_0[py]_0}{[Fe(III)]_0} \quad (49)$$

Thus the experimental determination of the rate constant k_e directly affords values of the intrinsic rate constant ratio k_1k_2/k_{-1}, since the quantities in brackets are constant under conditions of our kinetics experiments. This rate constant ratio is designated as k_H, i.e.,

$$k_H = k_1k_2/k_{-1} = Kk_2 \quad (50)$$

and the values listed in the second column of Table IX for PMT oxidations with the pyridine and 2,6-lutidine bases. The constant K in Eq. 50

Table X. Comparative Kinetics Data for the Reactions of Hexamethylbenzene and *p*-Methoxytoluene Cation Radicals toward Pyridine and 2,6-Lutidine.

Methylarene cation radical	*Base*	k_2^L/k_2^P	$k_2(H)/k_2(D)$[a]
CH_3O–(+)–CH_3	py	2.2	2.7±0.3
	lut		5.6±0.3
(+)–CH_3	py	2.2	3.6±0.3
	lut		3.9±0.3

[a]The first two entries are obtained from the data in Table IX by dividing the numbers in the second by the third column. The last two entries are taken from ref. 42.

represents the electron-transfer preequilibrium step in Eq. 46 and it is independently evaluated by the standard free energy change, i.e.,

$$K = k_1/k_{-1} = \exp(\mathcal{F}\,[E^0_{\mathrm{Fe}} - E^0_{\mathrm{Ar}}]\,/\,RT) \tag{51}$$

where E^0_{Fe} and E^0_{Ar} represent the standard reduction potentials for the $Fe(phen)_3^{3+}$ /$Fe(phen)_3^{2+}$ and the [$PMT^{\cdot+}$/[PMT] redox couples, respectively. The other terms have their usual significance (vide supra). To minimize the importance of the preequilibrium redox process, let us consider the significance of the rate constant k_H without any recourse to the reliance on the value of $(E^0_{\mathrm{Fe}} - E^0_{\mathrm{Ar}})$.[61] Thus it follows from Eq. 50 that the relative values of k_H for 2,6-lutidine and pyridine in Table IX actually represent the ratio of deprotonation rate constants since the redox preequilibrium constant K is common with both bases, i.e.,

$$k_H^L/k_H^P = k_2^L/k_2^P \tag{52}$$

where the superscripts L and P refer to 2,6-lutidine and pyridine, respectively. The relative rates of deprotonation are listed in Table X. In other words, 2,6-lutidine is more than twice as effective as pyridine in the followup step in Eq. 47. Likewise, the kinetic isotope effect reported as

k_H/k_D in column 4 of Table X represents only that for the deprotonation step in Eq. 47, i.e.,

$$k_H/k_D = k_2(\mathrm{H}) / k_2(\mathrm{D}) \tag{53}$$

The conclusion expressed in Eq. 53 leads to the question as to why the observed kinetic isotope effect of the followup reaction in Eq. 47 with 2,6-lutidine is almost *twice* as large as that with pyridine. The difference in basicity of 2,6-lutidine and pyridine can be ruled out as a cause of this discrepancy for the following reason. The kinetic study[42] of the hexamethylbenzene cation radical [$HMB^{\cdot+}$], which is similar to [$PMT^{\cdot+}$] indicates that there is no ambiguity as to the role of the pyridine, namely, it functions exclusively as a base for benzylic proton removal, i.e.,

$$[\mathrm{HMB}^{\cdot+}] + \mathrm{B} \xrightarrow{k'_2} \mathrm{C_6(CH_3)_5CH_2}\cdot + \mathrm{BH}^+ \tag{54}$$

It is noteworthy that proton transfer from [$HMB^{\cdot+}$] in Eq. 54 shows *no* difference in kinetic isotope effects between pyridine and 2,6-lutidine. The rate constants relevant to Eqs. 47 and 53 are retabulated in Table X to facilitate a direct comparison of the behavior of [$HMP^{\cdot+}$] and [$PMT^{\cdot+}$] toward pyridine and lutidine bases. Indeed the results in column 3 indicate that the relative rates of reaction of both cation radicals with pyridine and 2,6-lutidine are the *same*, which led to the conclusion that the rate measurements and the kinetic isotope effects for [$PMT^{\cdot+}$] do not refer completely to the same activation processes. In other words, the values of the kinetic isotope effects may not pertain only to reactions involving rate-limiting proton transfer. The key to the explanation of the difference in kinetic isotope effects for [$PMT^{\cdot+}$] with pyridine and 2,6-lutidine lies in the observation of two types of products **I** and **II** arising from side chain and nuclear substitution, respectively. If a kinetic isotope effect of ~6 is taken for the deprotonation rate constant k'_2 of [$PMT^{\cdot+}$] by pyridine, i.e.,

$$[\mathrm{CH_3OC_6H_4CH_3}]^{\cdot+} + \mathrm{B} \xrightarrow{k'_2} \mathrm{BH}^+ + \mathrm{CH_3OC_6H_4CH_2}\cdot \xrightarrow[\text{[-e]}]{\text{fast}} \longrightarrow \mathrm{I} \tag{55}$$

the observed value of 2.7 would imply that ~30% of the products result from benzylic proton removal, according to Eq. 55 (compare with Eq. 54), and the remainder from a process showing no kinetic isotope effect. Indeed the observation of only a 12% yield of the side chain substitution product **Ia**, accords with this formulation within the error limits of analysis. The remainder leading to nuclear substitution (see the formation of **IIa** in 70% yield in Table VIII) must then occur by a process in which proton transfer is fast. A mechanism involving the rate-limiting, nucleophilic addition of pyridine to [PMT$^{\bullet+}$] with a second-order rate constant k_2'' is consistent with this formulation, i.e.,

$$\text{PMT}^{\bullet+} + \text{Py} \xrightarrow{k_2''} [\text{Py}^+,\text{H-adduct radical}] \xrightarrow[\text{[-H}^+\text{]}]{\text{fast}} \longrightarrow \text{II} \quad (56)$$

Under these circumstances the observed rate constant k_2 is actually a composite [i.e., $k_2' + k_2''$], and pyridine serves a dual capacity as a base and a nucleophile in the followup step in Eq. 47. By contrast, 2,6-lutidine primarily functions only as a Brønsted base for deprotonation of the carbon acid, since no products of nuclear substitution could be observed. In this instance, the observed kinetic isotope effect truly reflects the deprotonation process in Eq. 55.

The formation of the *N*-arylpyridinium ion **IIa** resulting from nuclear substitution of [PMT$^{\bullet+}$] is unexpected since there are only a limited number of reports of such products in the extant literature. For example, the pyridination of anthracenes and 1,4-dimethoxybenzene, which have no methyl substituents, has been observed on anodic oxidation.[62] On the other hand, the cation radical of 9,10-dimethylanthracene was prone to deprotonation (i.e., benzylic substitution) rather than nucleophilic addition to the aromatic ring when a series of isomeric lutidines was employed as bases. 9-Phenylanthracene shows a preference toward deprotonation rather than nucleophilic attack at a cationic tertiary carbon center (addition of a second equivalent of nucleophile), depending on the steric requirements of the base. Such results identify the dichotomy that exists between nucleophilicity and basicity, as discussed by Eberson and Parker, insofar as the reactions of aromatic cation radicals are concerned. Perhaps a better comparison might be made for this competition by considering oxidative acetoxylation and oxidative cyanation with acetate and cyanide as the base (nucleophile). Thus the oxidative acetoxylation of *p*-methoxytoluene by either chemical or anodic oxidation results almost exclusively in benzylic products derived from side chain deprotonation.[60] On the other hand, anodic cyanation of PMT in acetonitrile results in nuclear substitution to afford predominant amounts of 6-methoxy-3-tolunitrile.[63] The favored *ortho* addition, which is analogous to the formation of **IIa** from pyridine (vide supra) is consistent with the calculated charge distribution in [PMT$^{\cdot+}$].[1]

The rate constants extracted from the kinetic data coupled with the product distribution lead to the evaluation of factors involved in the competition more quantitatively. Thus the overall rate constant k_H for [PMT$^{\cdot+}$] and pyridine in Table IX can be corrected for the competing nuclear substitution to afford the rate constant $k'_H = 5.01 \times 10^{-4}\ M^{-1}\ s^{-1}$ for deprotonation in Eq. 55. This value compares with that for [PMT$^{\cdot+}$] and 2,6-lutidine of $k_H = 3.74 \times 10^{-3} M^{-1} s^{-1}$, which also represents the deprotonation rate constant k'_H, since nuclear substitution is nil with this base. These rate constants for the deprotonation of [PMT$^{\cdot+}$] in Eq. 55 are related to those previously determined for [HMB$^{\cdot+}$] in Eq. 54.[42] Indeed the two sets of rate data can be compared quantitatively with the aid of the usual Brønsted relationship,[64] i.e.,

$$k^L_{2'}/k^P_{2'} = (K^L_a/K^P_a)^\alpha \tag{57}$$

where the superscripts L and P again refer to 2,6-lutidine and pyridine, respectively. Values of K^L_a and K^P_a, the acid dissociation constants of these pyridine bases, have been independently determined in acetonitrile.[65] It is important to note that the Brønsted slope evaluated from Eq. 57 is α = 0.28 for [PMT$^{\cdot+}$], which is the same as that ($\alpha = 0.26 \pm 0.02$) previously determined for HMB$^{\cdot+}$.[42] Therefore 2,6-lutidine and pyridine are differentiated in the deprotonation of either [PMT$^{\cdot+}$] or [HMB$^{\cdot+}$] only by *electronic factors* relating strictly to differences in their base strengths. By comparison, the rate constant for nucleophilic substitution of [PMT$^{\cdot+}$] by pyridine in Eq. 56 is $k''_H = 1.17 \times 10^{-3} M^{-1}\ s^{-1}$, whereas that for 2,6-lutidine is too small to measure despite its *higher* base strength. The relatively poor nucleophilic properties of 2,6-lutidine can be attributed to *steric effects* arising from a pair of *ortho*-methyl substituents that hinder the formation of the σ-adduct in Eq. 56. Such a formulation suggests a rather late transition state for nucleophilic substitution in which making the new C–N bond is important. By contrast, an early transition state pertains to the deprotonation of methylarene cation radicals by pyridine bases, since studies with [$HMB^{\cdot+}$] have established the driving force to lie in the exergonic region. Accordingly, it is noteworthy that the deprotonation rate constants are insensitive to steric effects of the pyridine bases, as expected for a process in which the making of the new bond between the proton and pyridine is minor.

6. DIRECT OBSERVATION OF ELECTROPHILIC AROMATIC SUBSTITUTION BY ELECTRON TRANSFER: PHOTOACTIVATION OF NUCLEAR NITRATION VIA CHARGE-TRANSFER COMPLEXES

The idea that charge transfer may play a key role in aromatic nitration with nitronium ion was first suggested in 1945 by Kenner,[66] who envisaged an initial step that "involves transference of a π-electron." Later Brown[67] postulated charge-transfer complexes as intermediates, and Nagakura[68] provided further theoretical support for one-electron transfer between an aromatic donor (ArH) and an electrophile such as NO_2^+. Despite notable elaborations by Pederson, Perrin, Eberson, and others, this formulation has not been widely accepted for nitration and related electrophilic aromatic substitutions.[69]

As broadly conceived, the seminal question focuses on the activation process(es) leading up to the well-established Wheland or σ-intermediate.[70] In the electron-transfer mechanism, the formation of the ion radical pair **III** is the distinctive feature, as summarized in Scheme 6:

$$ArH + NO_2^+ \underset{}{\overset{\text{fast}}{\rightleftharpoons}} [ArH,NO_2^+] \qquad \text{EDA Complex}$$

$$[ArH, NO_2^+] \xrightarrow{\text{slow}} [ArH^{\cdot+}, NO_2^{\cdot}] \quad \mathbf{III}$$

$$\mathbf{III} \xrightarrow{\text{fast}} Ar^{+}\!\begin{smallmatrix}H\\NO_2\end{smallmatrix} \qquad \text{Wheland Intermediate}$$

$$Ar^{+}\!\begin{smallmatrix}H\\NO_2\end{smallmatrix} \xrightarrow[B]{\text{fast}} ArNO_2 + HB^+$$

Scheme 6.

Accordingly, the properties and behavior of the intimate ion radical pair **III** are crucial to establishing its relationship with the numerous facets[71] of electrophilic aromatic nitration. For these reasons it is especially important to know whether **III** will actually lead to the appropriate Wheland intermediate, and in the amounts necessary to establish the isomer distributions commonly observed in aromatic nitrations. However, the independent proof of the ion radical pair **III** has not been forthcoming owing to its expectedly transitory character.

6.1. Charge-Transfer Nitration of Aromatic Donors

Picosecond time-resolved spectroscopy has defined the relevant photophysical and photochemical processes associated with the charge-transfer excitation of an arene complex such as anthracene with tetranitromethane.[72] As applied to benzenoid donors ArH, the formation of the pertinent ion radical pair by charge-transfer excitation is summarized below.

$$\mathrm{ArH} + \mathrm{C(NO_2)_4} \overset{k}{\rightleftharpoons} [\mathrm{ArH}, \mathrm{C(NO_2)_4}] \tag{58}$$

$$[\mathrm{ArH}, \mathrm{C(NO_2)_4}] \xrightarrow{h\nu_{CT}} [\mathrm{ArH}^{\cdot+}, \mathrm{C(NO_2)_4^{\cdot-}}] \tag{59}$$

$$[\mathrm{ArH}^{\cdot+}, \mathrm{C(NO_2)_4^{\cdot-}}] \xrightarrow{\text{fast}} \underset{\textbf{IV}}{[\mathrm{ArH}^{\cdot+}, \mathrm{NO_2^{\cdot}}, \mathrm{C(NO_2)_3^{-}}]} \tag{60}$$

Scheme 7.

All the experimental observations with various benzenoid donors and tetranitromethane indeed coincide with the formulation in Scheme 7. Thus the exposure of ArH to a nitrating agent such as TNM as in Figure 14 leads immediately to the EDA complex in Eq. 58. It is singularly noteworthy that the charge-transfer spectrum of the aromatic EDA complex with TNM is not fundamentally distinguished from the CT spectra of other common nitrating agents shown in Figure 14. Moreover all of these EDA binary complexes are present in low steady-state concentrations owing to the limited magnitude of K as measured by the Benesi–Hildebrand method. Activation of the EDA complex by the specific irradiation of the CT band results in a photoinduced electron transfer in accord with Mulliken theory.[73] Thus the irreversible fragmentation following the electron attachment to TNM leads to the ion radical pair **IV** in Eq. 60 (see Figure 15). The measured quantum yield of $\Phi \cong 0.5$ is similar to that ($\Phi \cong 0.7$) obtained for anthracene. Such high quantum yields relate directly to the efficiency of ion radical pair produc-

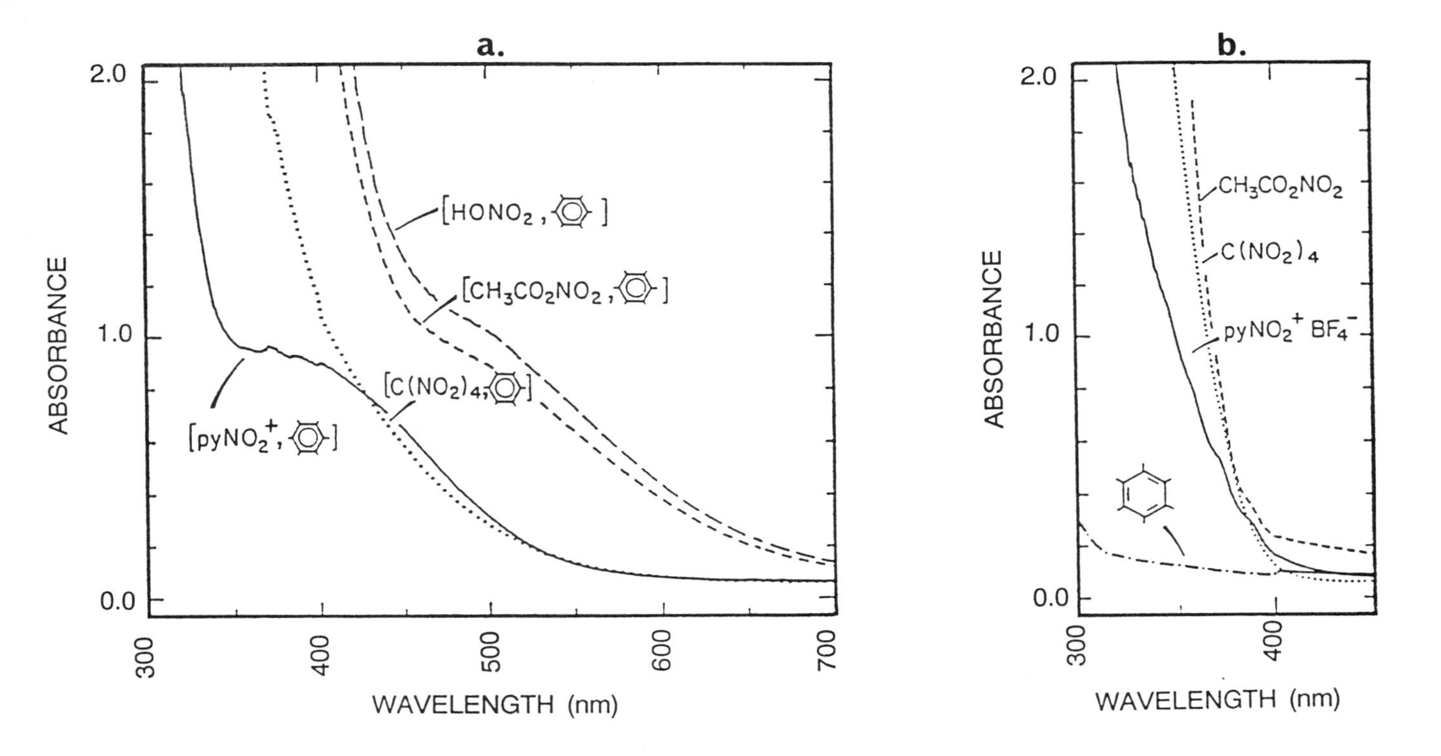

Figure 14. (a) Comparative Charge-Transfer Spectra of Hexamethylbenzene EDA Complexes with Various Nitrating Agents as Indicated. (b) Absorption Spectra of the Uncomplexed Donor and Acceptors.

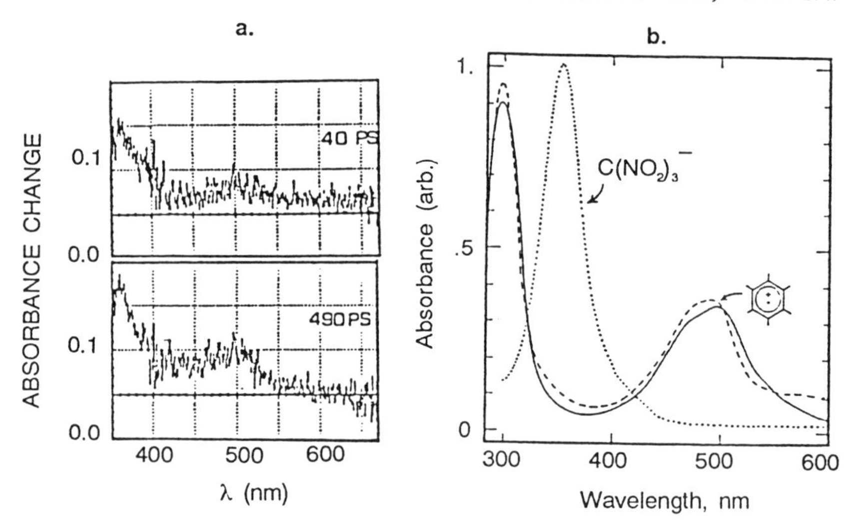

Figure 15. (a)Time-resolved Picosecond Absorption Spectra at 40 and 490 ps Following the 532-nm Laser Pulse Excitation of the Hexamethylbenzene EDA Complex with Tetranitromethane. (b) Absorption spectra of $HMB^{\bullet+}$ Generated Spectroelectrochemically and of $C(NO_2)_3^-$ Obtained from the Tetrabutylammonium Salt.

tion in Eq. 60 relative to energy wastage by back electron transfer in Eq. 59. Moreover the short lifetime (< 3 ps) of $C(NO_2)_4^{\bullet-}$ ensures that $ArH^{\bullet+}$ and $NO_2^\bullet$ are born as an intimate ion radical pair, initially trapped within the solvent cage, since this time scale obviates any competition from diffusional proceses.[72]

Charge-transfer excitation thus provides the ion radical pair **IV** to approximate the intimate ion radical pair [$ArH^{\bullet+}$, $NO_2^\bullet$] for Scheme 6 in sufficient concentrations and in a discrete electronic state as well as geometric configuration. Coupled with the observation of the fast kinetics allowed by the use of laser-flash photolytic techniques, we now focus on the pathways by which the ion radical pair collapses to nitration products with two series of aromatic ethers.

6.2. Dimethoxybenzenes as the Aromatic Donors in Nitration

p-Dimethoxybenzene (DMB) is the prototypical electron-rich aromatic donor owing to its reduced oxidation potential of only 1.30 V vs SCE. Charge-transfer excitation of the 1:1 DMB complex with TNM proceeds quantitatively according to the stoichiometry,

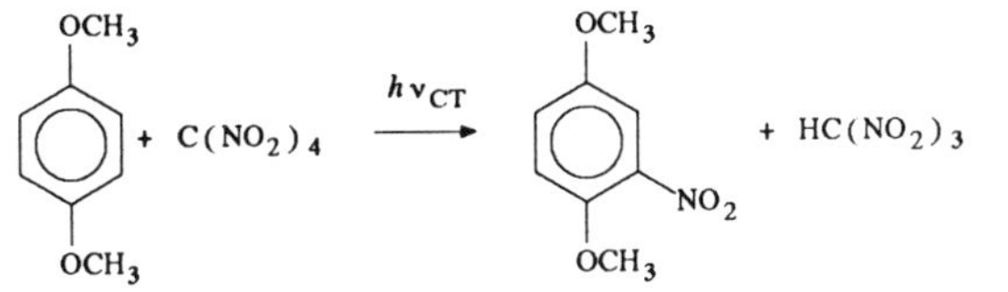

$$p\text{-}(CH_3O)_2C_6H_4 + C(NO_2)_4 \xrightarrow{h\nu_{CT}} 2\text{-}NO_2\text{-}1,4\text{-}(CH_3O)_2C_6H_3 + HC(NO_2)_3 \qquad (61)$$

The photochemical process is hereafter referred to as *charge-transfer nitration.* The excellent material balance obtained in charge-transfer nitration of DMB demands that the ion radical pair **IV** in Scheme 7 proceeds quantitatively to the nitration products according to the stoichiometry of Eq. 61.

$$\underset{\textbf{IV}}{[ArH^{\cdot+}, NO_2^{\cdot}, C(NO_2)_3^-]} \rightarrow ArNO_2 + HC(NO_2)_3 \qquad (62)$$

Such a transformation must occur spontaneously with no discrimination among the reactive intermediates to accord with the absence of a deuterium kinetic isotope effect. The latter is not consistent with the collapse of **IV** as an ion pair [$ArH^{\cdot+}$, $C(NO_2)_3^-$] by proton transfer to the very weakly basic trinitromethanide. Furthermore, the presence of extra trinitromethanide (deliberately added as the tetrabutylammonium salt TBAT) exerts essentially no influence on either the course or the kinetics. Accordingly the trinitromethanide is merely an innocuous bystander insofar as the conversion of the ion radical pair **IV** in Eq. 62 is concerned. It follows that the disappearance of the arene ion radical must be directly related to its interaction with $NO_2^{\cdot}$, i.e.,

$$\underset{\textbf{III}}{[ArH^{\cdot+}, NO_2^{\cdot}]} \rightarrow ArNO_2 + H^+ \qquad (63)$$

Indeed such cation radicals have been prepared from various arenes by other experimental methods, especially electrochemical oxidation.[74] The arene cation radicals related to $DMB^{\cdot+}$ are weak Brønsted acids, but they are highly susceptible to nuclear addition, *vide infra*, e.g.

$$\textbf{III} \longrightarrow Ar^{+}\!\!<^{H}_{NO_2} \qquad (64)$$

$$Ar^{+}\!\!<^{H}_{NO_2} \xrightarrow{k_H} ArNO_2 + H^+ \qquad (65)$$

Scheme 8.

The σ-adduct in Eq. 64 is the Wheland intermediate in electrophilic nitration, which is known to show no deuterium kinetic isotope effect for k_H on deprotonation in Eq. 65.[75] According to Scheme 8 the formation of the various isomeric Wheland intermediates will occur from the collapse of the ion radical pair in Eq. 64. Consequently the isomer distributions in the nitration products relate directly to the relative rates of addition to the various nuclear positions provided that it is irreversible and/or the adduct deprotonates rapidly. Thus the strong correlation observed between the spin densities at the various nuclear positions of $ArH^{\cdot+}$ and the isomeric product distribution in aromatic nitration[76] bears directly on the mechanism of such an ion radical pair collapse to the Wheland intermediate.

Although the Wheland intermediate in Scheme 8 has not been separately observed, the time-resolved spectral changes of the cation radical $ArH^{\cdot+}$ do provide insight as to how it is formed. Thus the relatively long lifetime of the rather stable arene cation radical $DMB^{\cdot+}$ is sufficient to allow diffusive separation of the ion radical pair **III** to $ArH^{\cdot+}$ and $NO_2^{\cdot}$ as essentially "free" species. The second-order process with the rate constant k_2 for the disappearance of $DMB^{\cdot+}$ then represents the "re"combination of these separated species to form the Wheland intermediate, i.e.,

$$\mathbf{III} \xrightarrow{<10^{-10}\mathrm{s}} ArH^{\cdot+} + NO_2^{\cdot} \qquad (66)$$

$$ArH^{\cdot+} + NO_2^{\cdot} \xrightarrow{k_2} Ar^{+}\begin{smallmatrix} H \\ NO_2 \end{smallmatrix} \qquad (67)$$

Scheme 9.

The strong solvent dependence of k_2 is largely attributed to the stabilization of $ArH^{\cdot+}$ by solvation, since it decreases with increasing solvent polarity in the order: hexane > benzene > dichloromethane. The slight negative salt effect on k_2 also accords with this conclusion. Since the common-ion salt TBAT has no effect on the charge-transfer nitration of DMB, the trinitromethanide is not a sufficiently strong nucleophile to intercept the associated cation radical $DMB^{\cdot+}$ in a process that would be tantamount to ion pair collapse of **IV**. However, it has been shown in electrophilic nitration that the Wheland intermediate can be efficiently captured by a built-in nucleophile, usually a carboxylic acid, and the nitration product isolated as an *ipso* adduct.[77]

Aromatic nitration via the outer-sphere ion radical pair **III** in Eq. 63 thus proceeds with high efficiency when it is induced by the charge-transfer excitation of arene–TNM complexes. Furthermore the yields and isomeric distributions among the nitration products from various dialkoxybenzenes are strikingly akin to those obtained under the more conventional electrophilic conditions.[78] One can conclude from these observations that intermediates leading to the conventional electrophilic nitration are similar to, if not the same as, those derived by charge-transfer nitration. Indeed the parallel behavior extends even to those arenes in which significant amounts of byproducts are formed. For example, the direct nitration of *m*-dimethoxybenzene is reported to produce 2,4-dimethoxynitrobenzene in only poor yields (~30 %).[79] In addition, an unusual blue coloration develops rapidly during the electrophilic nitration of *m*-dimethoxybenzene. The same intense blue color occurs with *m*-dimethoxybenzene and TNM, but only on the deliberate exposure of the EDA complex to CT irradiation. Among the dimethoxybenzenes, the *meta*-isomer is unique in that it is the only one to develop an intense (blue) coloration on electrophilic and/or charge-transfer nitration. The subsequent chromatography of the highly colored reaction mixture from electrophilic nitration yields significant amounts of dimethoxyphenyl dimers which are known to derive from the cation radical by arene coupling, i.e.,

$$1,3\text{-}(CH_3O)_2C_6H_4 \xrightarrow{m\text{-}(CH_3O)_2C_6H_4} \text{3,3',5,5'-}(CH_3O)_4\text{-biphenyl} \qquad (68)$$

and similar results are observed with 2,6-dimethoxytoluene and 1,3,5-trimethoxybenzene.

6.3. Haloanisoles as the Aromatic Donors in Nitration

The anisoles (XA) with 4-fluoro, chloro, and bromo substituents are substantially poorer electron donors as evidenced by their oxidation potentials E^0_{ox} that are ~500 mV more positive than that of DMB (*vide supra*). As a result, the corresponding cation radicals $XA^{•+}$ are significantly more susceptible than $DMB^{•+}$ to nucleophilic addition. The

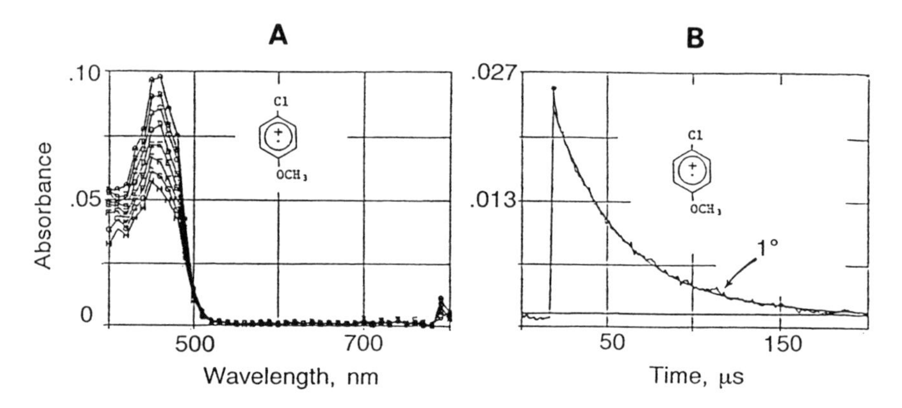

Figure 16. (A) Time-resolved Absorbtion Spectrum of the Donor Cation Radical Following the CT Excitation of the Tetranitromethane Complex with Arene (*p*-Chloroanisole). (B) First-order Decay of the Donor Cation Radical to the Wheland Intermediate in Aromatic Nitration.

latter can be circumvented in the CT nitration of haloanisoles by either the addition of neutral salt in dichloromethane or the use of acetonitrile as a polar medium. Most importantly the decay of the transient cation radical $XA^{\cdot+}$ formed by the CT excitation of the EDA consistently followed *first-order* kinetics. The clean first-order rate processes are applicable to the complete disappearance of $XA^{\cdot+}$ as established by the return of the absorbance to the baseline in Figure 16 for X = Cl. Since only CT nitration of XA occurs under these experimental conditions, the experimental first-order rate constant k_2 relates solely to the nuclear collapse of the ion radical pair

$$\left[\ \text{X-}C_6H_4^{\cdot+}\text{-}OCH_3\ ,\ \dot{N}O_2\ \right] \xrightarrow{k_2} \text{X-}C_6H_4^{+}(H)(NO_2)\text{-}OCH_3\ ,\ \text{etc.} \qquad (69)$$

Indeed the regiospecificity of such an ion radical pair collapse yields the isomeric mixture of nuclear nitration products that is essentially indistinguishable from that obtained under conventional electrophilic conditions (see Table XI). It is also worth noting that the byproducts from CT

Table XI. Aromatic Products from the Electrophilic (E) and Charge-Transfer (ET) Nitration of Anisole Derivatives.[a]

		Aromatic Products (mol %)					
Substituent in Anisole		*2-NO₂*	*3-NO₂*	*4-NO₂*	*5-NO₂*	*6-NO₂*	*Others*
None	ET	35	3	43			
	[E]	[31]	[2]	[67]			
2-CH_3	ET			68		32	
	[E]			[60]		[40]	
4-CH_3	ET	60					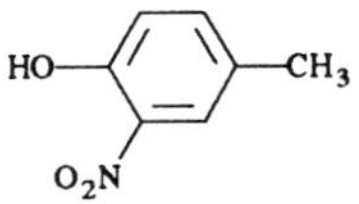 HO, O_2N, CH_3 (structure) 40
	[E]	80					20
4-F	ET	83					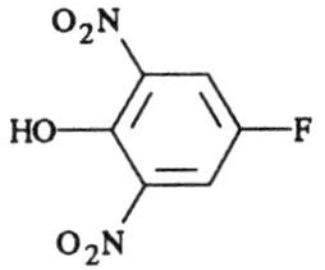 HO, O_2N, O_2N, F (structure) 14
	[E]	100					[0]

Table XI. (continued)

Substituent in Anisole		*Aromatic Products (mol %)*						
		2-NO_2	*3-NO_2*	*4-NO_2*	*5-NO_2*	*6-NO_2*	*Others*	
4-Br	ET	42					CH_3O–C_6H_4–NO_2	38
	[E]	48						[27]
							CH_3O–C_6H_3(Br)–Br	20
								[25]

[a]ET, electron transfer $C(NO_2)_4/h\nu_{CT}$, [E], electrophilic HNO_3/Ac_2O or HNO_3/H_2SO_4, 0°C.

nitration in Table XI are strongly reminiscent of the byproducts reported in electrophilic nitration of the anisoles with nitric acid. In particular, the demethylation of the methoxy group to afford nitrophenols, and the *trans*-bromination of 4-bromoanisole to afford a mixture of 4-nitroanisole and 2,4-dibromoanisole are both symptomatic of radical pair collapse at the *ipso* positions. These produce the σ-adducts, which are akin to the Wheland intermediates known to undergo such transformations (Eqs. 70 and 71).[80]

(70)

(71)

6.4. Structural Variation in the Kinetics of Ion Radical Pair Collapse

The kinetics of the collapse of the ion radical pair [$ArH^{\bullet+}$, $NO_2^{\bullet}$] from the representative arenes is summarized in Table XII. The decay of the spectral transients for nitration in Eq. 64 is a reflection of the stability of the aromatic cation radical. For example, the cation radical from *p*-methylanisole decays by second-order kinetics similar to the kinetic behavior of the long-lived cation radical from *p*-methoxyanisole. The large difference in the rates of diffusive combination with $NO_2^{\bullet}$ in Table XII (see column 4) corresponds to their relative stabilities as measured by $\Delta E^0 \equiv 8.5$ kcal mol^{-1} of the parent arenes (column 1). There is a further, larger gap of $\Delta E^0 \equiv$ 10.4 kmol mol^{-1} that separates the stabilities of the cation radicals of *p*-methylanisole and *p*-fluoroanisole, the least reactive haloanisole. Strikingly, every member of the family of *p*-haloanisole cation radicals reacts with $NO_2^{\bullet}$ by first-order kinetics. This decay pattern strongly suggests that the CT nitration occurs by the cage collapse of the geminate radical pair [$ArH^{\bullet+}$,$NO_2^{\bullet}$] prior to diffusive separation, except when the anisole cation is a relatively stabilized species such as those with *p*-methyl and *p*-methoxy substituents.

Table XII. Kinetics of the Charge-Transfer Nitration of Substituted Anisoles.[a]

$E°$ V vs SCE[a]	4-X–C_6H_4–OCH_3 *X*	*Kinetics order*[b]	*Rate constant*[c]
1.30	Methoxy	2°	1.0×10^4
1.67	Methyl	2°	2.5×10^5
2.12	Fluoro	1°	1.9×10^4
2.00	Chloro	1°	2.4×10^4
1.78	Bromo	1°	3.7×10^4

[a]In acetonitrile containing 0.1 *M* TBAP at 25°C.
[b]2° and 1° represent second- and first-order decays, respectively.
[c]In units of $A^{-1}s^{-1}$ for second-order and s^{-1} for first-order kinetics.

6.5. Decay Kinetics of Ion Radical Pairs as Reactive Intermediates

Since the charge-transfer complexes of NO_2^+ and arenes are too transient to exchange directly, we examined the cage collapse of the related geminate radical pair [$ArH^{•+}$,$NO^•$] since a nitrosonium salt like a nitronium salt can serve effectively either as an oxidant or as an electrophile toward different aromatic substrates.[81] Indeed NO^+ forms intensely colored charge-transfer complexes with a wide variety of common arenes (ArH), i.e.,

$$ArH + NO^+ \overset{K}{\rightleftharpoons} [ArH, NO^+] \tag{72}$$

For example, benzene, toluene, xylenes, and mesitylene generate yellow to orange vivid hues when added to colorless solutions of $NO^+PF_6^-$ in acetonitrile. Analogously, the more electron-rich durene, pentamethylbenzene, hexamethylbenzene, and naphthalene afford dark red solutions when exposed to NO^+. These charge-transfer colors are sufficiently persistent to allow single crystals of various arene CT complexes with NO^+ to be isolated for structural elucidation by X-ray crystallography.

Although $NO^+PF_6^-$ is insoluble in dichloromethane, it dissolves readily when an arene donor is present. These highly colored solutions on standing at –20°C deposit crystals of the CT complexes. In this manner, the 1:1 arene complexes [ArH, $NO^+PF_6^-$] are isolated with ArH = mesitylene, durene, pentamethylbenzene, and hexamethylbenzene. The ORTEP diagram from the X-ray crystallography of the mesitylene complex presented below[83] accords with those previously isolated by

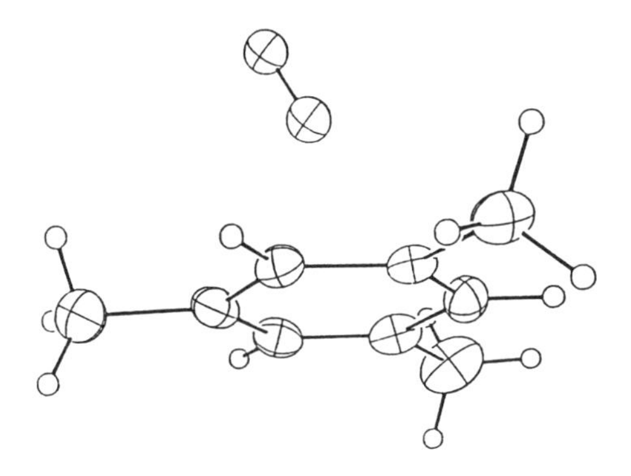

Brownstein and co-workers from liquid sulfur dioxide.[82] Indeed the relevant CT interaction clearly derives from the centrosymmetric (η^6) structure of the arene–NO^+ pair reminiscent of the other aromatic EDA structures presented earlier.[11] However, the NO^+ complexes are unusual in two important ways. First, the formation constant K in Table XIII strongly depends on the donor strength (i.e., ionization potential) for the arene—increasing dramatically from 0.5 M^{-1} with benzene to 31,000 M^{-1} with hexamethylbenzene (HMB). Second, the frequency of the N–O stretching band in the infrared spectrum decreases markedly from ν_{NO} = 2037 cm^{-1} in the toluene complex to ν_{NO} = 1880 cm^{-1} in the hexamethyl-

Table XIII. Formation Constants of Arene Charge-Transfer Complexes with NO^+.[a]

Arene (10^{-3} M)		IP (eV)	$NO^+BF_4^-$ (10^{-3} M)	λ_{CT} (nm)	K (M^{-1})	ε_{CT} ($M^{-1}cm^{-1}$)
Benzene	(41–580)[b]	9.23	9.8	346	0.46	780
Toluene	(45–75)	8.82	6.0	342	5.0	400
Mesitylene	(3.2–23)	8.42	0.81	345	56	2080
1,3,5-Tri-*t*-butylbenzene	(3.3–19)	8.19	0.92	349	34	1730
Pentamethylbenzene	(2.7–15)	7.92	0.20	337	5100	2400
Hexamethylbenzene	(0.15–17)	7.71	0.20	337	31000	3000

[a]In acetonitrile at 25°C.
[b]Lower–upper concentration range.

Table XIV. Infrared Spectra of Arene CT Complexes with $NO^+PF_6^-$.

$[ArH,NO^+PF_6^-]$ ArH	Solid (cm^{-1}) NO^+ (fwhm)[b]		Δ[c]	Solution (cm^{-1})[a] NO^+ (fwhm)	
Hexamethylbenzene	1899	(152)	441	1880	(49)
Pentamethylbenzene	1927	(196)	413	1904	(59)
Durene	1986	(191)	355	1929	(60)
Mesitylene	2016	(157)	324	1967	(93)
p-Xylene	*d*			1998[e]	(92)
o-Xylene	*d*			2000[e]	(97)
Toluene	2042[f]	(180)	298	2037[e]	(91)

[a]In dichloromethane.
[b]Full width half maximum.
[c]Shift from free $NO^+PF_6^-$ at 2340 cm^{-1}.
[d]Not measured.
[e]In nitromethane.
[f]Ground mixture of toluene with $NOPF_6$.

benzene complex (Table XIV). Such a large change in ν_{NO} parallels the difference between the uncomplexed acceptor [$\nu(NO^+) = 2280\ cm^{-1}$] and free nitric oxide [$\nu(NO) = 1876\ cm^{-1}$].

The unusually pronounced dependence of both the formation constant K and N–O stretching frequency of the aromatic EDA complexes with NO^+ is illustrated in Figure 17. Indeed the strong correlations with the aromatic donor strength as evaluated by the ionization potential point to a sizable change in the charge-transfer component in the ground state of these complexes.[84] In particular, the value of ν_{NO} in the hexamethylbenzene (HMB) complex, which is essentially the same as that of free NO, suggests complete electron transfer in the ground state.[85]

According to Mulliken theory, the ground state of weak EDA complexes (with K typically < 10 M^{-1}) can be described as

$$\psi = a\,\psi_{D,A} + b\,\psi_{D^+A^-} \tag{73}$$

with the coefficients $a >> b$ to denote a minor contribution from the charge-transfer state.[86] The complete reversal on CT excitation generates an excited state with a large (total) contribution from the charge-transfer state, i.e.,

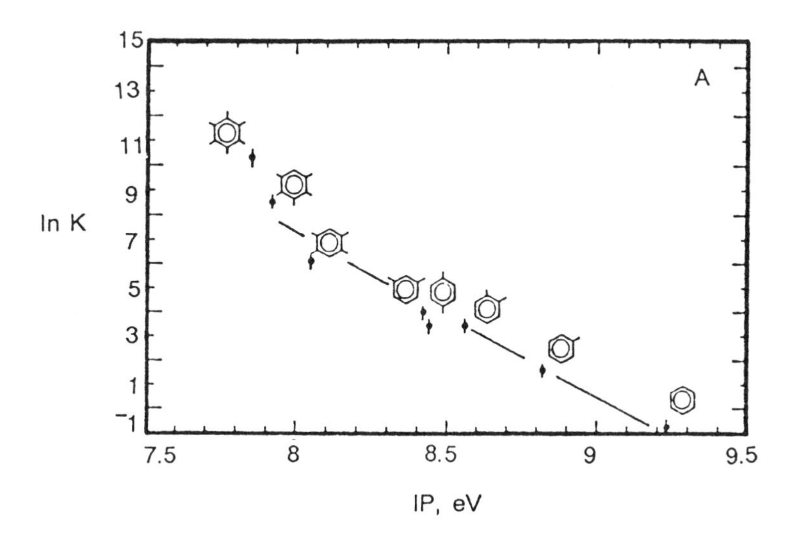

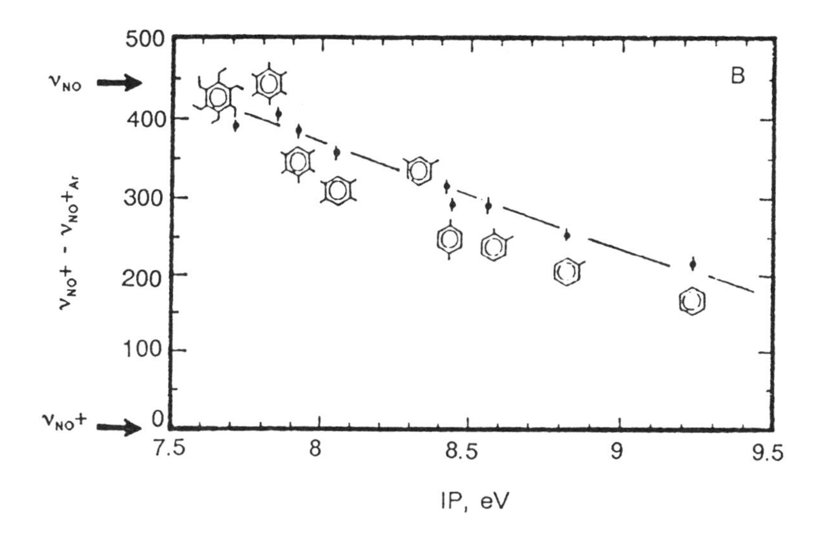

Figure 17. Variations of (A) the Formation Constant K and (B) the N–O Stretching Frequencies ($\nu_{NO^+} - \nu_{NO^+,Ar}$) in the Infrared Spectra of 1:1 EDA Complexes of NO^+ and Various Arenes as Indicated.

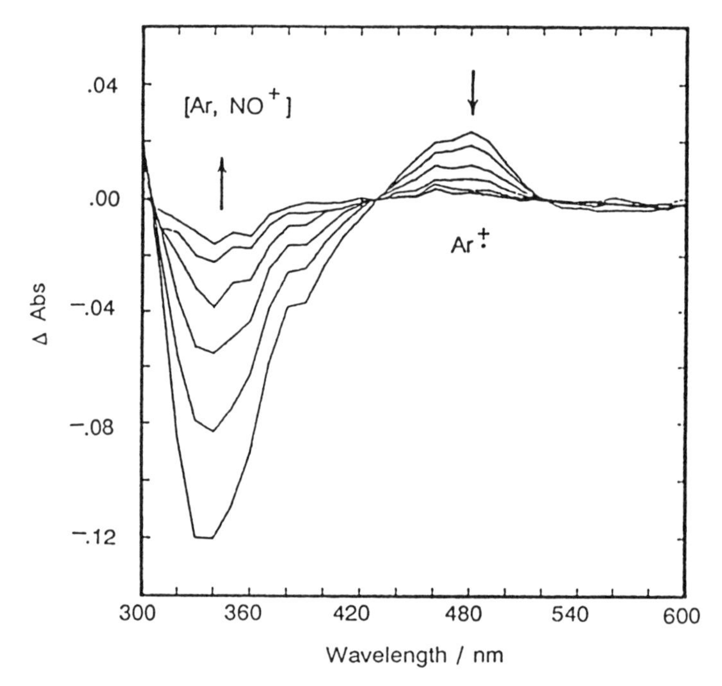

Figure 18. Typical Time-resolved Absorption Spectrum Following the CT Excitation of Nitrosonium EDA Complexes with Arene (Hexamethylbenzene) Showing the Bleaching of CT Absorption and Growth of the Donor Cation Radical ($HMB^{\cdot+}$).

$$\psi^{*} = b\,\psi_{D,A} - a\,\psi_{D^{+}A^{-}} \tag{74}$$

Indeed, the time-resolved spectroscopy studies presented in Sections 6.2 and 6.3 relate directly to the formation of ion radicals as CT transients from such weak EDA complexes. By an analogous reasoning, the ground state of strong EDA complexes such as $HMB\text{-}NO^{+}$ ($K > 10^{4}\,M^{-1}$ and ν_{NO} the same as that of free NO) may be described essentially as an EDA complex of the cation radical with NO, i.e., [$HMB^{\cdot+}$, NO]. The slope of the correlation in Figure 17A indicates a sharp trend toward the benzene complex in which the ground state can be largely represented by the no-bond structure (with $a >> b$ in eq. 73).

The excitation of the charge-transfer band of the aromatic EDA complexes with NO^{+} is carried out at $\lambda = 355$ nm using a 30-ns laser pulse. The time-resolved spectrum in Figure 18 shows the typical bleach-

Table XV. Time-Resolved Spectroscopy of the CT Activation of Aromatic EDA Complexes with NO^+.[a]

Substituent(s) in Benzene	λ *($ArH^{\bullet+}$) (nm)*	k_{-1} *(10^6 s^{-1})*
None	430	2.7
Cl	470	1.1
Br	540	1.5
I	680	1.4
Me	430	1.2
1,4-Me_2	440	1.4
1,2,4-Me_3	455	1.0
1,2,4,5-Me_4	460	0.66
Me_5	485	0.30
Me_6	495	0.15

[a]In dichloromethane solution by CT excitation at 355 nm.

ing of the CT absorption band and the appearance of the aromatic cation radical with λ_{max} ~500 nm,[87] i.e.,

$$[ArH,NO^+] \underset{k_{-1}}{\overset{h\nu_{CT}}{\rightleftharpoons}} [ArH^{\bullet+}, NO] \qquad (75)$$

Since no photochemistry occurs upon the prolonged irradiation of the CT absorption band, the subsequent first-order decay of the $ArH^{\bullet+}$ absorption clearly relates to back electron transfer (k_{-1}). It is significant that the trend in k_{-1} (Table XV) accords with the expected increase in driving force for back electron transfer in proceeding from hexamethylbenzene ($IP = 7.71$ eV) to benzene ($IP = 9.23$ eV). Moreover it is interesting to note that the magnitudes of these rate constants are in line with those reported in Table XII for the ion radical pair [$ArH^{\bullet+}$, NO_2] in charge-transfer nitration. Be that as it may, the time-resolved spectral studies support the notion that ion radical pairs can be held together by CT interactions sufficient to retard their diffusive separation.

6.6. Comments on the Mechanism of Aromatic Nitration

The foregoing results underscore the strong similarity between electrophilic and charge-transfer nitration of aromatic ethers, with regard

$$[ArH,\ NO_2^+] \underset{k_{-1}}{\overset{k_1}{\rightleftharpoons}} [ArH^{\bullet+}, NO_2^{\bullet}] \qquad (76)$$

$$[ArH^{\bullet+}, NO_2^{\bullet}] \xrightarrow{k_c} Ar^{+}\!\!\left\langle\begin{matrix} H \\ NO_2 \end{matrix}\right. \qquad (77)$$

$$[ArH^{\bullet+}, NO_2^{\bullet}] \xrightarrow{k_d} ArH^{\bullet+} + NO_2^{\bullet} \qquad (78)$$

$$Ar^{+}\!\!\left\langle\begin{matrix} H \\ NO_2 \end{matrix}\right. \xrightarrow{\text{fast}} ArNO_2 + H^+ \qquad (79)$$

Scheme 10.

to both the nitration products and the aryl byproducts. Since the latter is symptomatic of arene cation radicals, the parallelism does extend to some reactive intermediates in common. As such, the most *economical* formulation for electrophilic nitration would also include a pathway that is common to charge-transfer nitration, namely, via the ion radical pair **III**, as presented in Scheme 6. This mechanism differs from the conventional formulation in which the [ArH, NO_2^+] pair is directly converted to the Wheland intermediate in a single step, rather than via the ion radical pair **III**. Previous studies have established the strong similarity in the activation barriers (i.e., energetics) for these two processes.[76] Accordingly the problem can be considered in an alternative framework, namely, the lifetime of the ion radical pair. At one extreme of a very short lifetime, the inner-sphere ion radical pair is tantamount to the transition state for the concerted one-step process. At the other extreme of a very long lifetime, the outer-sphere ion radical pair is manifested by its diffusive separation. Scheme 10 presents this construct in a kinetics context.[8]

1. The extent to which $k_d >> k_c$ allows side reactions to compete as a result of the diffusive separation of $ArH^{\bullet+}$ and $NO_2^{\bullet}$. The experiments with aromatic diethers clearly belong in the latter category since the evolution of nitration products takes place in the microsecond/millisecond time regime. Nonetheless the diffusive recombination of $ArH^{\bullet+}$ and $NO_2^{\bullet}$ pairs with the second-order rate constant k_2 can occur in Eq. 67 with very high efficiency. The competition from diffusion is represented by the intermolecular trapping of $ArH^{\bullet+}$ in Eq. 68.

2. The efficient collapse of the ion radical pair to the Wheland intermediate is represented by $k_c >> k_d$. Such a direct collapse of the ion radical pair has its counterpart in CT nitration by the observed first-order decay of the spectral transients derived from the 4-haloanisole donors (compare Figure 16). Indeed the clean first-order rate constants listed in Table XII for the ion radical pair collapse in Eq. 69 relate to the rate constant k_c in Eq. 77 of Scheme 10. However, the magnitude of $k_2 \sim 10^4$ s^{-1} is at least six orders of magnitude slower than that expected from the diffusional correlation times of $10^{-10} - 10^{-11}$ s; the measured first-order rate constants in Table XV indicate that such ion radical pairs can persist for prolonged periods.

7. ION PAIR VERSUS RADICAL PAIR ANNIHILATION OF ARENE CATION RADICALS: SALT AND SOLVENT EFFECTS

Cation radicals from electron-rich arenes are sufficiently persistent to allow their physical properties to be examined in detail.[88] On the other hand, $ArH^{\cdot+}$ from other less endowed arenes can be quite transient, which limits their study to more indirect methods, especially with regard to their reactivity. Among these properties, the ambivalent character of the delocalized paramagnetic ion is of particular interest since its cationic charge offers coulombic attraction for anions ($A:^-$), whereas its unpaired electron promotes homolytic reactions with free radicals ($R^\cdot$).[5] Both the ion pair and radical pair interactions can lead to aromatic substitution when they collapse at a nuclear position, e.g.,

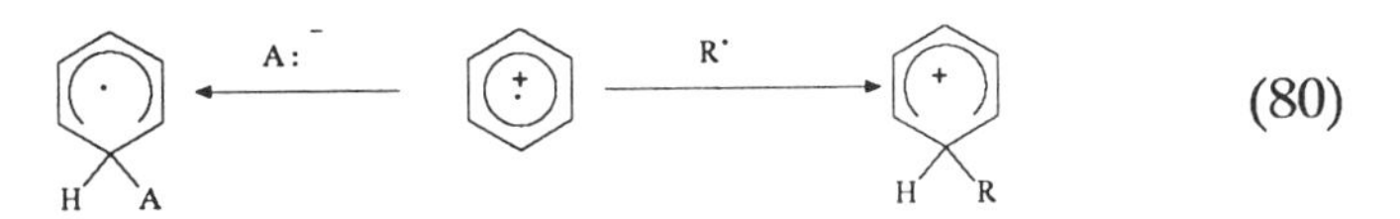

(80)

Thus the cationic adduct represents the Wheland intermediate in electrophilic aromatic substitution, and the radical adduct is the cyclohexadienyl intermediate relevant to homolytic aromatic substitution.[89]

The competition between the ion pair and the radical pair annihilation in Eq. 80 can be examined directly by the charge-transfer photochemistry described in Scheme 7. For these studies the electron-rich dialkoxybenzenes in Eq. 61 were replaced with the anisole and the haloanisoles in Table XVI.[72] Most noteworthy for the anisoles in Table XVI is the

Table XVI. Charge-Transfer Trinitromethylation of Anisole and Its Derivatives in Dichloromethane.[a]

Aromatic Donor	*2-C(NO2)3*	*Other Products*	*Material Balance*
$C_6H_5OCH_3$	40[b]	2-NO_2 anisole (22%) 4-NO_2 anisole (28%)	90
2-$CH_3C_6H_4OCH_3$	60[c]	2-Me-4-NO_2 anisole (10%) 2-Me-6-NO_2 anisole (6%)	76
3-$H_3CC_6H_4OCH_3$	62[d]	3-Me-NO_2 anisoles[e]	—
4-$H_3CC_6H_4OCH_3$	95	4-Me-2-NO_2 anisoles (5%)	100
4-$FC_6H_4OCH_3$	72	4-F-2-NO_2 anisole (9%) 4-F-2,6-$(NO_2)_2$ anisole (6%)	87
4-$ClC_6H_4OCH_3$	67	4-Chloro-2-NO_2 anisole (7%)	74
4-$BrC_6H_4OCH_3$	73	4-Bromo-2-NO_2 anisole (7%) 4-NO_2 anisole (6%) 2,4-Br_2 anisole (5%)	91

[a]By CT excitation of the EDA complexes from 0.10 *M* ArH and 0.8 *M* TNM at λ>425 nm at 25°C.

[b]4-$(O_2N)_3C$-anisole.

[c]2-Me-4-$(O_2N)_3C$-anisole.

[d]3-Me-4-$(O_2N)_3C$-anisole.

[e]Mixture of 2,4 and 6-NO_2-3-Me-anisole.

pronounced modulation of the competition by solvent polarity and by added salts. Thus the efficient collapse of ion pairs in nonpolar solvents (Table XVI) is completely subverted to radical pair collapse by a simple change to polar solvents (Table XVII), i.e.,

$$\text{ArH(A)}^{\bullet} \xleftarrow[\text{solvent}]{\text{nonpolar}} [\text{ArH}^{\bullet+}, A:^-, R^{\bullet}] \xrightarrow[\text{solvent}]{\text{polar}} \text{ArH(R)}^{+} \quad (81)$$

where $A:^-$ and $R^{\bullet}$ represent $C(NO_2)_3^-$ and $NO_2^{\bullet}$ respectively, in this study. Strikingly, the mere presence of low concentrations of innocuous salts such as the quaternary ammonium perchlorates (e.g., TBAP in Table XVIII) in nonpolar solvents effects the same diversion from ion pair collapse to radical pair collapse. i.e.,

$$\text{ArH(A)}^{\bullet} \xleftarrow{\text{no salt}} [\text{ArH}^{\bullet+}, A:^-, R^{\bullet}] \xrightarrow{\text{salt}} \text{ArH(R)}^{+} \quad (82)$$

Comparison of the spectral transients from the CT irradiation of the EDA complex in dichloromethane and in acetonitrile (Figure 19) indicates that the same species are formed. In other words, solvent polarity plays little or no role in the critical formation of the ion/radical triad **IV**. Moreover the measured quantum yields Φ show little variation with solvent polarity. The value of $\Phi = 0.2$ for 4-methyl anisole compares with $\Phi = 0.5$ and 0.7 for dialkoxybenzenes and anthracenes, respectively,[72,90] measured under the same experimental conditions. These high quantum yields relate directly to the efficiency of triad production in Eq. 60 relative to energy wastage by back electron transfer in Eq. 59. Indeed the trend $\Phi = 0.2$, 0.5, and 0.7 accords with the expected relative rates of back electron transfer based on the oxidation potentials of $E^{\circ} = 1.67, 1.30$, and 1.23 V for 4-methylanisole, 1,4-dimethoxybenzene and anthracene, respectively.

The short lifetimes of less than 3 ps for $C(NO_2)_4^{\bullet-}$ (as established in the anthracene studies)[72] indicates that the ion/radical triad **IV** is initially trapped within the solvent cage since this timescale precludes any competition from diffusional separation. Thus the vertical excitation of the EDA complex according to Scheme 7 ensures that $ArH^{\bullet+}$ and $C(NO_2)_3^-$ are born as a discrete, intimate ion pair in which the mean separation is

Table XVII. Charge-Transfer Nitration of Anisole and Derivatives in Acetonitrile Solutions.[a]

Aromatic Donors	*2-NO2*	*3-NO2*	*4-NO2*	*5-NO2*	*6-NO2*	*Others*	
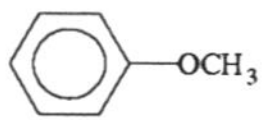 OCH_3	35	3	43				
CH_3, OCH_3			68				
H_3C, OCH_3	16		54				
H_3C–, –OCH_3	60					HO–, –CH_3, O_2N	40
F–, –OCH_3	83					O_2N, HO–, –F, O_2N	14

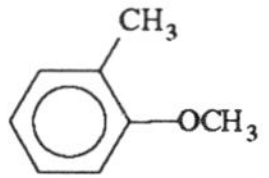

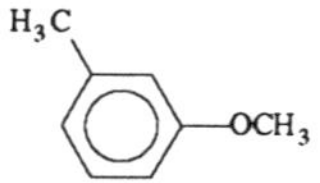

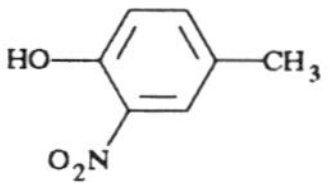

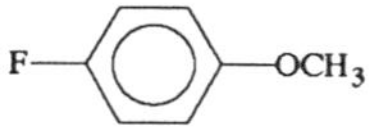

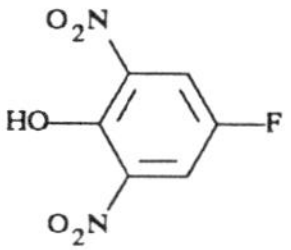

Table XVII. (Continued)

Aromatic Donors	*2-NO_2*	*3-NO_2*	*4-NO_2*	*5-NO_2*	*6-NO_2*	*Others*
Cl–C_6H_4–OCH_3	100					
Br–C_6H_4–OCH_3	42					CH_3O–C_6H_4–NO_2 38; CH_3O–C_6H_3(Br)–Br 20

[a]See Table XVI for conditions.

Table XVIII. Salt Effects on Charge Transfer Substitutions.[a]

Solvent	*ArH*[b] (*M*)	*TNM* (*M*)	*Salt*[c] (*M*)	*Time*[d] (*h*)	*Products*: 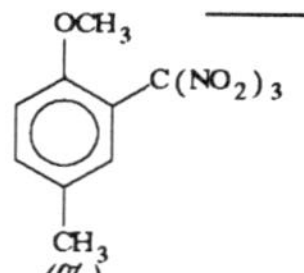 4-methyl-2-(trinitromethyl)anisole (%)	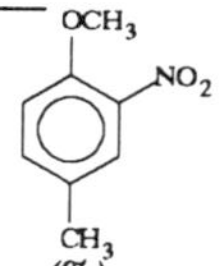 4-methyl-2-nitroanisole (%)	*Isolated Yield* (%)
CH_2Cl_2	0.06	0.66		5	95	5	70
CH_2Cl_2	0.06	0.83	TBAP (0.2)	5	0	100[e]	
CH_2Cl_2	0.10	0.55	TBAP (0.01)	6	0	100[f]	
CH_2Cl_2	0.06	2.22	TBAT (0.01)	6.5	76	24	
MeCN	0.06	1.67		6.5	5	95[g,j]	
MeCN	0.13	1.39	TBAP (0.2)	6	0	100[h,k]	
MeCN	0.06	1.39	TBAT (0.01)[l]	7	10	90[i]	
n-C_6H_{14}	0.06	0.28		5	85	15	70
C_6H_6	0.03	0.55		5	85	15	72
C_6H_6	0.06	0.55	TBAP (0.01)[m]	3.5	75	25	
C_6H_6	0.06	1.50	TBAT (0.01)[m]	6.5	100	0	

[a]From CT irradiation of 3 mL of solution with $\lambda > 425$ nm.

[b]4-Methylanisole.

[c]Tetra-*n*-butylammonium trinitromethide (TBAT) and perchlorate (TBAP).

[d–k]Duration of irradiation. Includes 4-methyl-2-nitrophenol in [e]35%, [f]39%, [g]14%, [h]6%, [i]13% yield. Includes 4-methyl-2,6-dinitrophenol in [j]15%, [k]39% yield.

[l]Irradiation with $\lambda > 480$ nm

[m]Saturated solution.

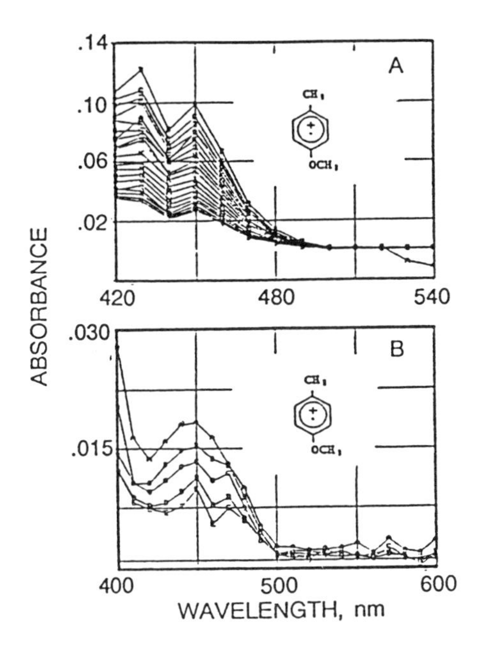

Figure 19. Spectral Transients from the CT Irradiation with λ = 532 nm of 0.1 *M* Methylanisole and 0.8 *M* TNM in (A) Acetonitrile and (B) Dichloromethane.

largely dictated by the geometry extant in the precursor EDA complex.[91] Likewise, $ArH^{\bullet+}$ and $NO_2^{\bullet}$ exist in **IV** as the geminate radical pair akin to those formed in bond homolysis.[92] The pathways by which the caged triplets suffer ion pair and radical pair collapse, and ultimately lead to aromatic alkylation and nitration, respectively, can be considered as follows. The excellent material balances obtained in Tables XVI and XVII demonstrate that the caged triad **IV** proceeds quantitatively to the aromatic products by one of two stoichiometries, viz.

$$[ArH^{\bullet+}, C(NO_2)_3^-, NO_2^{\bullet}] \text{ (IV)} \longrightarrow ArC(NO_2)_3 + HNO_2 \quad (83)$$

$$[ArH^{\bullet+}, C(NO_2)_3^-, NO_2^{\bullet}] \text{ (IV)} \longrightarrow ArNO_2 + HC(NO_2)_3 \quad (84)$$

Under conditions in which alkylation and nitration are observed simultaneously, the combination of Eqs. 83 and 84 accounts for the complete material balance for CT substitution. Since aromatic alkylation clearly derives from the coupling of the aromatic cation with the trinitromethide anion, the pathway for Eq. 83 is designated as *ion-pair collapse*. Analogously, the pathway to aromatic nitration in Eq. 84 is identified as *radical pair collapse*. The absence of a deuterium kinetic isotope effect in either aromatic alkylation or nitration (vide supra) effectively precludes the loss of the aromatic hydrogen as H$^{\bullet}$ or H^{+} in a rate-limiting step. Accordingly a two-step process can be formulated in which the rate of aromatic alkylation is determined by the rate constant k_{ip} from prior ionic addition, i.e.

$$[\mathrm{ArH}^{\bullet+}, \mathrm{C(NO_2)_3^-}, \mathrm{NO_2^{\bullet}}] \xrightarrow{k_{ip}} \overset{\bullet}{\mathrm{Ar}}\begin{smallmatrix}\mathrm{H}\\ \mathrm{C(NO_2)_3}\end{smallmatrix} + \mathrm{NO_2^{\bullet}} \quad (85)$$
IV

$$\overset{\bullet}{\mathrm{Ar}}\begin{smallmatrix}\mathrm{H}\\ \mathrm{C(NO_2)_3}\end{smallmatrix} + \mathrm{NO_2^{\bullet}} \xrightarrow{\text{fast}} \mathrm{ArC(NO_2)_3} + \mathrm{HNO_2} \quad (86)$$

Scheme 11.

The σ-adduct formed in Eq. 85 is analogous to the intermediates in homolytic alkylation.[89,93] As such, the hydrogen atom transfer in Eq. 86 is expected to be facile, and this conclusion accords with the absence of a deuterium isotope effect in CT alkylation. By an analogous formulation, the rate of aromatic nitration is determined by the rate constant k_{rp} from *homolytic* addition, i.e.,

$$[\mathrm{ArH}^{\bullet+}, \mathrm{C(NO_2)_3^-}, \mathrm{NO_2^{\bullet}}] \xrightarrow{k_{rp}} \mathrm{Ar}^{+}\begin{smallmatrix}\mathrm{H}\\ \mathrm{NO_2}\end{smallmatrix} + \mathrm{C(NO_2)_3^-} \quad (87)$$
IV

$$\mathrm{Ar}^{+}\begin{smallmatrix}\mathrm{H}\\ \mathrm{NO_2}\end{smallmatrix} + \mathrm{C(NO_2)_3^-} \xrightarrow{\text{fast}} \mathrm{ArNO_2} + \mathrm{HC(NO_2)_3} \quad (88)$$

Scheme 12.

The σ-adduct formed in Eq. 87 is the Wheland intermediate in electrophilic nitration.[70] As such it relates directly to the classical studies of Melander,[75] who first established the rapid, irreversible loss of proton from the Wheland intermediate.

The dichotomy between aromatic alkylation and nitration as described in Schemes 11 and 12 provides the basis for understanding the extraordinary solvent and salt effects on the behavior of transient aromatic cation radicals. In order to carry out the mechanistic analysis, the earlier studies in which the dynamics of ion pairs similar to that in **IV** are reconsidered.[74] Of particular concern is the segregation of the ion and radical fragments in **IV**, especially with regard to diffusive separation. This is accomplished by first considering the dynamics of ion pair separation. Such a priority is dictated by the dominance of the inherent Coulombic interactions in enthalpic considerations, and the large driving force provided by the neutralization of the opposite charges in the formation of adduct in Eq. 85. The earlier time-resolved spectroscopic studies[72] established the existence of three successive rate profiles of ion pairs. These were ascribed to (1) the fast first-order decay (k_1) of the first-formed, intimate ion pair [$ArH^{•+}$, $C(NO_2)_3^-$] to the "loose" or solvent-separated ion pair [$ArH^{•+}//C(NO_2)_3^-$], which subsequently suffer (2) another first-order, but slower decay (k'_1) to the free ions $ArH^{•+}$ and $C(NO_2)_3^-$, and finally (3) the second-order diffusive combination (k_2) of the separated ions. The ion pair dynamics summarized below

$$\underset{\sigma\text{-adduct}}{\mathbf{II}} \rightleftharpoons \underset{\text{"contact" ion pair}}{[ArH^{•+}, C(NO_2)_3^-]} \overset{h\nu_{CT}\downarrow}{} \rightleftharpoons \underset{\text{"loose" ion pair}}{[ArH^{•+} // C(NO_2)_3^-]} \rightleftharpoons \underset{\text{"free" ions}}{ArH^{•+} + C(NO_2)_3^-} \quad (89)$$

Scheme 13.

drew on the pioneering studies of Winstein and co-workers for solvolytic processes.[94] As such, the quantitative treatments of the "special salt" and common ion effects, as represented by TBAP and TBAT, respectively, provide important support for the substantiation of this formulation. Accordingly the time-resolved spectroscopic studies help to delineate the role of both the solvent polarity and added salts in CT alkylation as well as in CT nitration. Before doing so however, it is important to emphasize that the clean stoichiometries in Tables XVI and XVII for Eqs. 85 and 87, respectively, allow us to examine the rates of these processes by spectrally following the disappearance of only the aromatic cation radical. Thus the stoichiometry for alkylation (Eq. 83) and nitration (Eq. 84) demands the concomitant disappearance of an equivalent amount of the anion $C(NO_2)_3^-$ and the radical $NO_2^•$, respectively.

Ion pair annihilation in Eq. 85 is favored by nonpolar solvents, as indicated by the high yields of the alkylation products in Table XVI. The decay pattern of the spectral transient provides interesting insight as to how the ion pair collapse occurs, since there are large changes with even minor variations in the solvent and with small amounts of added salt. In hydrocarbon solvents (n-hexane and benzene) the rate profile for Eq. 85 (Scheme 11) follows first-order kinetics, which reflect the ion-pair collapse from the geminate ions originally present in **IV**. However, in slightly more polar solvents such as dichloromethane, the rate profile for Eq. 85 follows a different kinetics behavior. Indeed the magnitudes of the second-order rate constants in dichloromethane are reminiscent of those obtained earlier for the combination of the separate, free ions.[74] Such a diverse behavior parallels the increased lifetimes of ions in the more polar medium such as dichloromethane relative to hexane. Under these circumstances, the enhanced diffusive separation of the ion pair will be reflected in the change in the kinetics from first-order to second-order behavior. In terms of Scheme 13, the change in kinetics from hydrocarbon solvents to dichloromethane represents a change in mechanism for alkylation from the collapse of the intimate or solvent-separated ion pair [$ArH^{\cdot+}//C(NO_2)_3^-$] to diffusive combination of $ArH^{\cdot+}$ and $C(NO_2)_3^-$. In accord with this conclusion, it is noteworthy that the series of first-order rate constants k_1 in hydrocarbon solvents follow essentially the same trend as the second-order rate constants k_2 in dichloromethane. Both rate constants (i.e., log k_1 and log k_2) reflect the energetics of ion pair collapse in Eq. 85. The relatively large gap between the reactivity of the p-methylanisole and p-fluoroanisole cations parallels the difference in the oxidation potential of the parent anisoles, i.e., E^0 = 1.67 and 2.12 V, respectively. As such, it largely represents the difference in the stabilities of the anisole cation radicals. By comparison the highly stabilized cation radical from the CT excitation of p-methoxyanisole with E^0 = 1.30 V is incapable of ion pair collapse with trinitromethide anion. Thus the decay kinetics and the products lead to the conclusion that the primary distinction between benzene (or hexane) and dichloromethane as solvents for ion pair annihilation arises from differences in dielectric constants that allow increased separation of the ions in the more polar solvent.

The effects of added salt on the decay kinetics of the ion pair can be interpreted in terms of the "special salt" and "common-ion" effects on the ion pair equilibria in Scheme 13. For example, the presence of the innocuous salt tetrabutylammoniumperchlorate (TBAP) in benzene alters the decay from first-order to second-order kinetics. In terms of Scheme 13, the effect of added TBAP is to exchange the active ion pair with an inactive ion pair (i.p.), i.e.

$$[ArH^{\cdot+}, C(NO_2)_3^-] + [Bu_4N^{\cdot+}, ClO_4^-] \rightleftharpoons$$
active i.p.

$$[ArH^+, ClO_4^-] + [Bu_4N^+, C(NO_2)_3^-] \qquad (90)$$
inactive i.p.

which is tantamount to ion pair separation by the "special salt" effect. This change is also reflected in decreased amounts of CT alkylation (see Table XVIII, entry 10). The fact that CT alkylation is not completely wiped out by TBAP indicates that the ion pair exchange in Eq. 90 is not thorough in the highly nonpolar hydrocarbon media. Conversely, ion pair separation is completely inhibited by the presence of TBAT—the decay process maintaining first-order kinetics as a result of common ion suppression, i.e.,

$$\underset{\text{(free)}}{ArH^{\cdot+}} + [Bu_4N^+, C(NO_2)_3^-] \rightleftharpoons \underset{\text{(active i.p.)}}{[ArH^{\cdot+}, C(NO_2)_3^-]} + Bu_4N^+ \qquad (91)$$

As expected, this is reflected in exclusive CT alkylation (Table XVIII, entry 11).

The effects of added salts on the decay profile in dichloromethane are qualitatively the same as those described above in hexane. Thus the second-order kinetics observed in this more polar solvent persists in the presence of added TBAP. However, the product analysis in Table XVIII (entry 2) shows the complete absence of CT alkylation. The second-order kinetics must therefore be associated with CT nitration, even at low levels (0.01 *M*) of TBAP (entry 3). Furthermore, the presence of the common-ion salt TBAT is inadequate to completely counter this large positive salt effect (see Table XVIII, entry 4).

The increased efficiency with which added salts affect the ionic ion pair equilibria (Scheme 13) in dichloromethane relative to hexane can be attributed to the slightly greater polar properties of the solvent (vide supra). This change presumably acts on the ion pair exchange such as that in Eq. 90, which effectively separates the cationic $ArH^{\cdot+}$ from the anionic trinitromethide. The aggregation of ion pairs particularly in the highly nonpolar (and poorly solvating) hydrocarbons[95] accounts for the decreased efficiency of such ion pair exchange.

In the absence of competition for ion pair collapse as described in the preceding section, the CT excitation of the EDA complex leads only to aromatic nitration. This is particularly evident with electron-rich arenes such as *p*-methoxyanisole, which afford highly persistent cation radicals incapable of ion pair annihilation with trinitromethide anion in all solvents, including hexane. As a result, quantitative yields of nitration products are always obtained with a variety of dialkoxybenzenes on the CT irradiation of their EDA complexes with tetranitromethane. With the more reactive aromatic cation radicals, the competition from ion pair separation becomes increasingly more important only in highly polar solvents such as acetonitrile. Indeed the dramatic change in the course of CT substitution from aromatic alkylation to nitration parallels the change in solvent from dichloro methane to acetonitrile (compare Tables XVI and XVII).

According to Scheme 12, the mechanism of CT nitration depends on the particular pathway by which $ArH^{\bullet+}$ encounters $NO_2^{\bullet}$. Indeed, the decay pattern of the spectral transients for nitration in acetonitrile is strikingly dependent on the stability of the aromatic cation radicals as measured by the one-electron oxidation potentials E^0 of the arene, i.e., The direct comparison of E^0 in Table XII focuses on the substantial effect of substituents (X) on the inherent stability of the aromatic cation radicals.

$$X\text{-}C_6H_4\text{-}OCH_3 \overset{E^\circ}{\rightleftharpoons} [X\text{-}C_6H_4\text{-}OCH_3]^{\bullet+} + e \quad (92)$$

The decay of the spectral transients for nitration in Eq. 87 (Scheme 12) is a reflection of the stability of the aromatic cation radical. For example, the cation radical from *p*-methylanisole decays by second-order kinetics in acetonitrile similar to the kinetic behavior of the long-lived cation radical from *p*-methoxyanisole. The large difference in the rates of diffusive combination with $NO_2^{\bullet}$ in Table XII (see column 4) corresponds to their relative stabilities as measured by $\Delta E^0 = 8.5$ kcal mol^{-1} of the parent arenes (column 1).

There is a further, larger gap of $\Delta E^0 = 10.4$ kcal mol^{-1}, which separates the stabilities of the cation radicals of *p*-methylanisole and *p*-fluoroanisole, the least reactive haloanisole. Strikingly, every member of the family of *p*-haloanisole cation radicals reacts with $NO_2^{\bullet}$ by first-order kinetics. This decay pattern strongly suggests that the CT nitration occurs by the cage collapse of the geminate radical pair [$ArH^{\bullet+}$, $NO_2^{\bullet}$] prior to diffusive separation, except when the anisole cation is a relatively stabilized species such as those with *p*-methyl and *p*-methoxy substituents. Such a mechanism for nitration implies that polar solvents act selectively to preferentially separate the anion from the cation and $NO_2^{\bullet}$ in the geminate triad **IV**. This conclusion is also consistent with the marked effect of added innocuous salt TBAP in the less polar solvent dichloromethane. It is noteworthy that the second-order rate constant for nitration under these conditions is the same as that obtained in acetonitrile.

The isomer distributions in CT alkylation and nitration are established in Schemes 11 and 12 during ion pair and radical pair collapse according to Eqs. 85 and 87, respectively. Since the driving forces for both of these processes are likely to be exergonic, the transition states for the formation of the relevant aromatic σ-adducts will bear a strong resemblance to the aromatic cation radical.[96] As such, the charge and electron density in $ArH^{\bullet+}$ will be an important factor in determining the positional selectivity in the aromatic ring.[97] As a first approximation, the regiospecificity in ion pair and radical pair collapse should be the same, since the site of highest positive charge in $ArH^{\bullet+}$ will also be the site of highest electron spin density. The latter are determined from EPR spectra, which show that these locations in the cation radicals of anisole and its derivatives are concentrated at the *ortho* and *para* positions relative to the methoxy group, e.g.,[5]

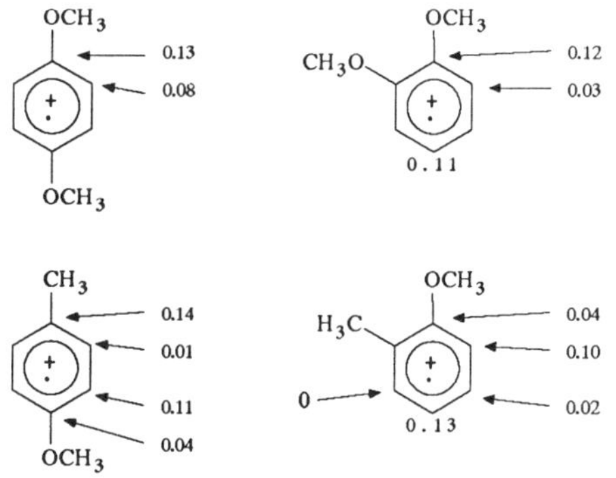

Examination of the products in Tables XVI and XVII indeed shows that CT nitration and alkylation both occur at sites *ortho* and *para* to the methoxy group. However, the absence of the *ortho* isomers during CT alkylation of anisole as well as 2- and 3-methylanisoles in Table XVII indicates that the introduction of the bulky trinitromethyl group is somewhat subject to steric hindrance. Otherwise, the collapse of the ion pair

[$ArH^{\cdot +}$, $C(NO_2)_3^-$] and the radical pair [$ArH^{\cdot +}$, $NO_2^{\cdot}$] to the σ-adducts leads to the same positional selectivity in CT alkylation and nitration. The σ-adduct arising from the radical pair collapse is the Wheland intermediate in electrophilic nitration. Thus it is worth noting that the byproducts from CT nitration in Table XVII are strongly reminiscent of the byproducts reported in electrophilic nitration of the anisoles with nitric acid. In particular, the demethylation of the methoxy group to afford nitrophenols, and the *trans*-bromination of 4-bromoanisole to afford a mixture 4-nitroanisole and 2,4-dibromoanisole are both symptomatic of radical pair collapse at the *ipso* positions. These produce the σ-adducts, which are akin to the Wheland intermediates known to undergo such transformations as described in Eqs. 70 and 71.

It is important to emphasize that the comparison of the behavior of various *p*-substituted anisoles shows that ion pair collapse is strongly dependent on the stability of the cation radical, being nonexistent with the highly stabilized *p*-methoxyanisole cation radical. The rates of ion pair collapse of 4-substituted anisole cations increase in the order: methyl < fluoro < chloro < bromo in line with the oxidation potentials E^0 of the parent anisole. Moreover the rates of radical pair collapse also increase in the same order, but with a change in the decay behavior from second-order kinetics for *p*-methylanisole to first-order kinetics for all the haloanisoles. The latter points to the importance of a geminate process for radical pair collapse from **IV** with reactive cation radicals. Clearly further studies are required to substantiate this interesting facet of solvent effects on the diffusive behavior of cations, anions, and neutral radicals. Nonetheless the complex solvent effect on the decay patterns of the ion/radical triad **IV** can be accounted for in a consistent manner if ion pair collapse in Eq. 69 is the preferred process, except for aromatic cation radicals, which are stabilized either by electron-donor substituents (e.g., *p*-methyl and *p*-methoxy) by solvation (e.g., acetonitrile) or by added salt (TBAP).

8. ELECTRON TRANSFER ACTIVATION IN THE THERMAL AND PHOTOCHEMICAL OSMYLATIONS OF AROMATIC EDA COMPLEXES WITH OSMIUM(VIII) TETROXIDE

The various facets of aromatic activation by electron transfer as presented in Sections 2–7 are coherently brought together in a typical electrophilic reaction involving osmium tetroxide. We hope the thorough treatment of this subject will encourage similar studies with other electrophiles.

Oxo-metals have drawn increased attention as viable oxidation catalysts for various types of oxygen–atom transfers to organic and biochemical substrates.[98] However, only scant mechanistic understanding exists of the oxidation pathways, certainly with respect to the nature of the activation barrier and the identification of the reactive intermediate(s). Among oxo-metals, osmium tetroxide is a particularly intriguing oxidant since it is known to rapidly oxidize various types of alkenes, but it nonetheless eschews the electron-rich aromatic hydrocarbons like benzene and naphthalene.[99] Such selectivities do not obviously derive from differences in the donor properties of the hydrocarbons since the oxidation (ionization) potentials of arenes are actually less than those of alkenes. The similarity in the electronic interactions of arenes and alkenes toward osmium tetroxide relates to the series of electron donor–acceptor (EDA) complexes formed with both types of hydrocarbons,[100] i.e.,

$$[Ar, OsO_4] \underset{}{\overset{Ar}{\rightleftharpoons}} OsO_4 \underset{}{\overset{Alk}{\rightleftharpoons}} [Alk, OsO_4] \qquad (93)$$

Common to both arenes and alkenes is the immediate appearance of similar colors that are diagnostic of charge-transfer (CT) absorptions arising from the electronic excitation ($h\nu_{CT}$) of the EDA complexes formed in Eq. 93. As such, the similarity in the color changes point to electronic interactions in the arene complex $[Ar, OsO_4]$ that mirror those extant in the alkene complex $[Alk, OsO_4]$.

The charge-transfer colors of the alkene EDA complexes are fleeting, and they are not usually observed owing to the rapid followup rate of osmylation.

$$>C=C< + OsO_4 \rightleftharpoons [Alk, OsO_4] \xrightarrow{fast} \text{(cyclic osmate ester) } OsO_2, \text{etc.} \qquad (94)$$

By contrast, simple (monocyclic) arenes do not afford thermal adducts with osmium tetroxide, benzene actually being a most desirable solvent for alkene hydroxylation. However, with some extended polynuclear aromatic hydrocarbons such as benzopyrene, dibenzanthracene, and cholanthrene, a thermal reaction does lead to multiple osmate adducts and finally to polyhydric alcohols.[101] Tricyclic aromatic hydrocarbons such as phenanthrene show intermediate reactivity with osmium tetroxide to afford (over several weeks) the 1:1 adduct.

$$\text{phenanthrene} + OsO_4 \rightleftharpoons [Ar, OsO_4] \longrightarrow \text{(9,10-dioxo-osmate ester, } OsO_2\text{)} \quad (95)$$

Osmylation in Eq. 95 occurs at the HOMO site of the arene in a manner analogous to that observed in alkene osmylations (Eq. 94). In this context, anthracene is a particularly noteworthy substrate since it is purported to afford an unusual 2:1 adduct by oxidative attack at a terminal ring in preference to the most reactive *meso* (9,10-) positions.[102]

The inextensible range of arene reactivities offers the unique opportunity to probe the mechanism of osmium tetroxide oxidations for four principal reasons. First, the EDA complexes in Eq. 93 relate alkenes directly to arenes via the oxo-metal interactions in the precursors relevant to oxidation. Second, the thermal osmylation of polynuclear arenes (see Eq. 95) has an exact counterpart in the photostimulated osmylations that are widely applicable to even such otherwise inactive arenes as benzene. Since this charge-transfer process is readily associated with excitation ($h\nu_{CT}$) of the EDA complex to the ion pair,

$$[Ar, OsO_4] \underset{k_{-1}}{\overset{h\nu_{CT}}{\rightleftharpoons}} [Ar^{\bullet+}, OsO_4^{\bullet-}] \xrightarrow{\text{fast}} \text{Adduct} \qquad (96)$$

it is hereafter referred to simply as charge-transfer osmylation. Third, the dual pathways of *thermal* and *charge-transfer* osmylation allow the regio- and stereochemistry for OsO_4 addition to be quantitatively compared, especially in the OsO_4 adducts of the polycyclic arenes: phenanthrene, anthracene, and naphthalene. Fourth, the activation process for CT osmylation can be unambiguously established by the application of time-resolved (picosecond) spectroscopy for direct observation of the reactive intermediates, as previously defined in other aromatic CT processes.[103] Accordingly, our initial task in this study is to establish the common CT character of the EDA complexes of OsO_4 with the arene series: benzene, naphthalene, and anthracene as well as the structural elucidation of their OsO_4 adducts by X-ray crystallographic and spectral analyses.

8.1. Aromatic EDA Complexes with Osmium(VIII) Tetroxide

A colorless solution of osmium tetroxide in hexane or dichloromethane on exposure to benzene turns yellow instantaneously.[104] With durene an orange coloration develops and a clear bright red solution results from hexamethylbenzene. The quantitative effects of the dramatic color changes are illustrated in Figure 20 by the spectral shifts of the electronic absorption bands that accompanies the variations in aromatic conjugation and substituents. The progressive bathochromic shift parallels the decrease in the arene ionization potentials (*IP*) in the order: benzene 9.23 eV, naphthalene 8.12 eV, anthracene 7.55 eV, much in the same manner as that observed with the tropylium acceptor earlier.[105] Such spectral behaviors are diagnostic of electron donor–acceptor complexes [Ar, OsO_4]. According to Mulliken, the new absorption bands derive from charge-transfer excitation with the energetics defined by $h\nu_{CT} = IP - E_A - w_p^*$ (vide supra), where E_A is the electron affinity of the OsO_4 acceptor and w_p^* is the dissociation energy of the CT excited ion pair state [$Ar^{\bullet+}$, $OsO_4^{\bar{}}$].

8.2. Thermal Osmylation of Naphthalene, Anthracene, and Phenanthrene

Benzene shows no signs of osmylation in the absence of light, as indicated by the persistence of the yellow color of the [C_6H_6, OsO_4] complex in hexane even on prolonged standing. On the other hand, the orange CT color of the phenanthrene complex [$C_{14}H_{10}$, OsO_4] slowly diminishes over a period of weeks, accompanied by the formation of a dark brown precipitate of the composition ($C_{14}H_{10}OsO_4$). Dissolution of the solid in pyridine yields the 1:1 adduct ($C_{14}H_{10}OsO_4py_2$) **P** as the sole product in very low conversion. Anthracene behaves similarly to afford the 2:1 adduct in 10% conversion only after 2 months. The thermal osmylation can be expedited in a purple solution of refluxing heptane (100°C) to effect a 68% conversion in 30 h. However, even at these relatively elevated temperatures, naphthalene is converted to the corresponding 2:1 adduct to only a limited extent. In every case, the dark brown primary adducts are easily collected from the reaction mixture as

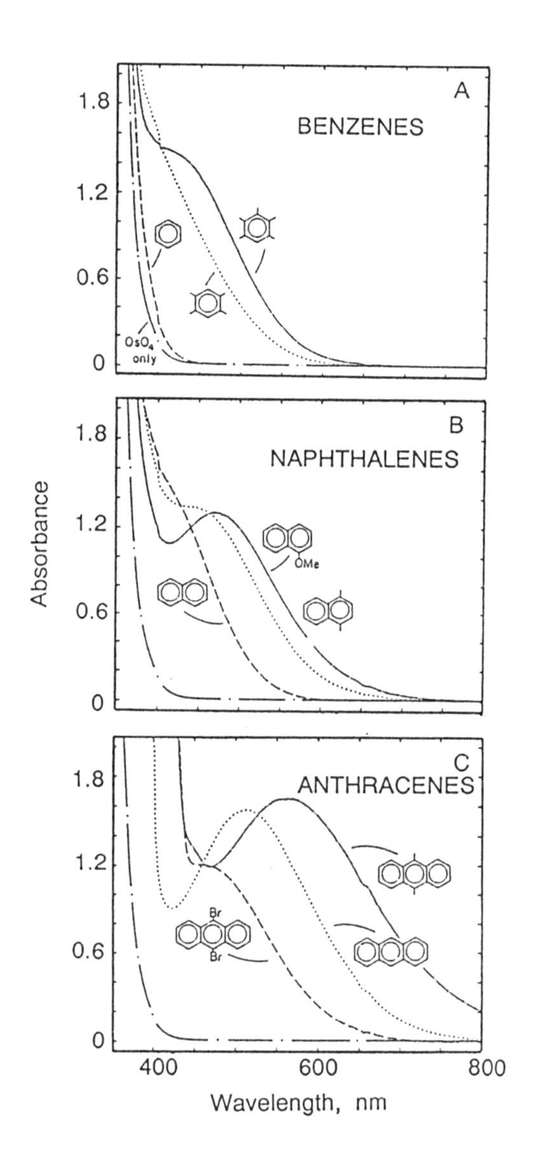

Figure 20. Typical Bathochromic Shifts of the CT Absorption Bands of Osmium Tetroxide EDA Complexes with Decreasing Ionization Potentials of Arene Donors: Benzenes (A) < Naphthalenes (B) < Anthracenes. (C).

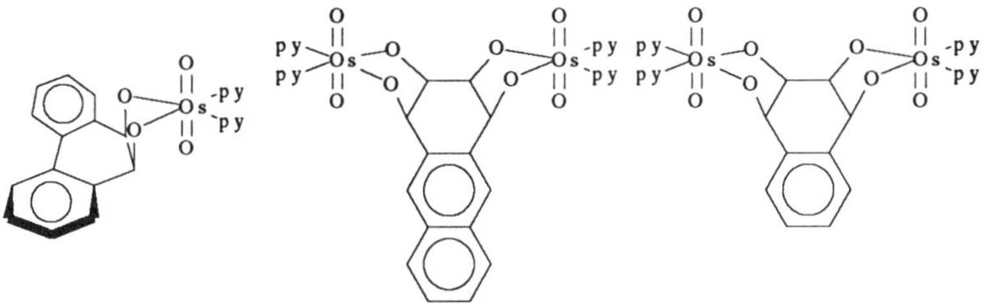

P A N

insoluble solids, and then immediately ligated with pyridine for structural characterization. Indeed the characteristic IR and ^{1}H NMR spectra of the anthracene, phenanthrene, and naphthalene adducts **A**, **P**, and **N**, respectively, allows the ready analysis of the osmylated adducts. Since these adducts are derived from the arenes with only OsO_4 present, the chemical transformation is hereinafter designated as the *direct thermal* or DT osmylation. For comparison, the same polynuclear arenes can be osmylated in the presence of promoter bases, typically pyridine. Under these conditions, the adducts **A**, **P**, and **N** are formed directly in the reaction mixture and at substantially increased rates of reaction, as previously established with the related family of alkene substrates. Such a procedure differs visually from the DT osmylation described above in that the charge-transfer colors are not observed as transients, owing to the preferential complexation of OsO_4 with pyridine. Accordingly, this *promoted thermal* or PT osmylation is to be distinguished by the enhanced reactivity of the pyridine complex relative to the free OsO_4 in the DT osmylation. The corresponding increase in the yields of adducts such as **A**, **P**, and **N** within a shorter span of reaction times is apparent from the comparison of the results of DT and PT osmylations.

8.3. Charge-Transfer Osmylation of Benzene, Naphthalene, and Anthracene

The various charge-transfer colors for the different arene complexes with OsO_4 are persistent for days. However, when the colored solutions are deliberately exposed to visible light with energy sufficient to excite only the charge-transfer band, they always deposit a highly insoluble, dark brown solid of the OsO_4 adducts obtained from the direct thermal osmylation of arenes (*vide supra*). Because this photoprocess must have arisen via the electronic excitation of the EDA complex, it is referred to hereafter as *charge-transfer* or CT osmylation for individual arenes. For example, the irradiation of the charge-transfer bands (see Figure 20) of the OsO_4 complexes with various benzenes, naphthalenes, and phenanthrene yield the same osmylated adducts such as **N** and **P** described above. Anthracene is unique in that it affords two entirely different types of products on the photoexcitation of the EDA complex [$C_{14}H_{10}$, OsO_4] in dichloromethane and hexane, despite only minor solvent effects on the charge-transfer bands. Irradiation of the purple solution of anthracene and OsO_4 in dichloromethane at $\lambda > 480$ nm yields the 2:1 adduct **A** together with its *syn* isomer as the sole products. On the other hand, irradiation of the same purple-colored solution but in hexane under otherwise identical conditions leads to a small amount of polymeric osmium dioxide $(OsO_2)_x$. Workup of the hexane solution yields anthraquinone as the major product contaminated with only traces (<1%) of the 2:1 adduct **A**. Interestingly, even higher yields of anthraquinone are obtained from 9-bromo, 9-nitro, and 9,10-dibromoanthracene when the CT osmylation is carried out in hexane. Such an accompanying loss of the electronegative substituents (X = Br, NO_2) probably occurs via osmylation at the *meso* (9,10-) positions followed by oxidative decomposition of the unstable adduct with the stoichiometry

$$\text{X-anthracene} + OsO_4 \xrightarrow[C_6H_{14}]{h\nu_{CT}} \text{adduct (X, O–}OsO_2\text{–O)} \longrightarrow \text{anthraquinone} + OsO_2 + HX \qquad (97)$$

8.4. Time-Resolved Spectra of Arene Cation Radicals in Charge-Transfer Osmylation

To identify the reactive intermediates in the charge-transfer excitation of arene–OsO_4 complexes, the time-resolved spectra are measured immediately following the application of a 30-ps pulse consisting of the second harmonic at 532 nm of a mode-locked Nd^{3+}:YAG laser. The wavelength of this excitation source corresponds to the maxima (or near maxima) of the charge-transfer absorption bands of the series of anthracene complexes with osmium tetroxide illustrated in Figure 20C. Accordingly, the time-resolved spectra from the anthracene-OsO_4 system relates directly to the CT osmylation since there is no ambiguity about either the adventitious local excitation of complexed (or uncomplexed) chromophores, or the photogeneration of intermediates that did not arise from the photoexcitation of the EDA complex. Indeed, intense absorptions are observed in the visible region between 700 and 800 nm from the excitation of the anthracene–OsO_4 complex, as shown in Figure 21A. This time-resolved absorption spectrum from anthracene is obtained in the time interval of ~30 ps following the application of the 532-nm laser pulse. Comparison with the steady-state absorption spectrum of the anthracene cation radical (see inset) generated by the spectroelectrochemical technique, thus establishes the identity of the charge-transfer transient. Similar time-resolved spectra of arene cation radicals are obtained from various anthracene and naphthalene EDA complexes despite the excitation of only the low-energy tails of the CT bands in Figure 20 with the 532-nm laser pulse. The evolution of the anthracene cation radical is followed by measuring the absorbance change at λ_{max} = 742 nm upon the charge-transfer excitation of the EDA complex with a single laser shot of ~10 mJ. The time evolution of the absorbance shown in Figure 21B includes the initial onset for ~20 ps as a result of the rise time of the 30-ps (fwhm) laser pulse. The first-order plot of the decay portion is shown in the inset to the figure. Decay curves similar to those shown in Figure 21 are also observed for the disappearances of the cation radicals derived from all of the other arene–OsO_4 complexes. In each case, the highest concentrations are obtained of the arene cation radical, and all the decay kinetics correspond to first-order processes. The magnitudes of the rate constant are applicable to the complete disappearance of $Ar^{\bullet+}$ as indicated by the return of the cation-radical absorbances to the baseline.

8.5. Common Features in Thermal and Charge-Transfer Osmylations

The [Ar, OsO_4] complexes are involved as the common precursors in the oxidative addition of osmium tetroxide to various arenes by the three independent procedures designated as direct thermal (DT), promoted

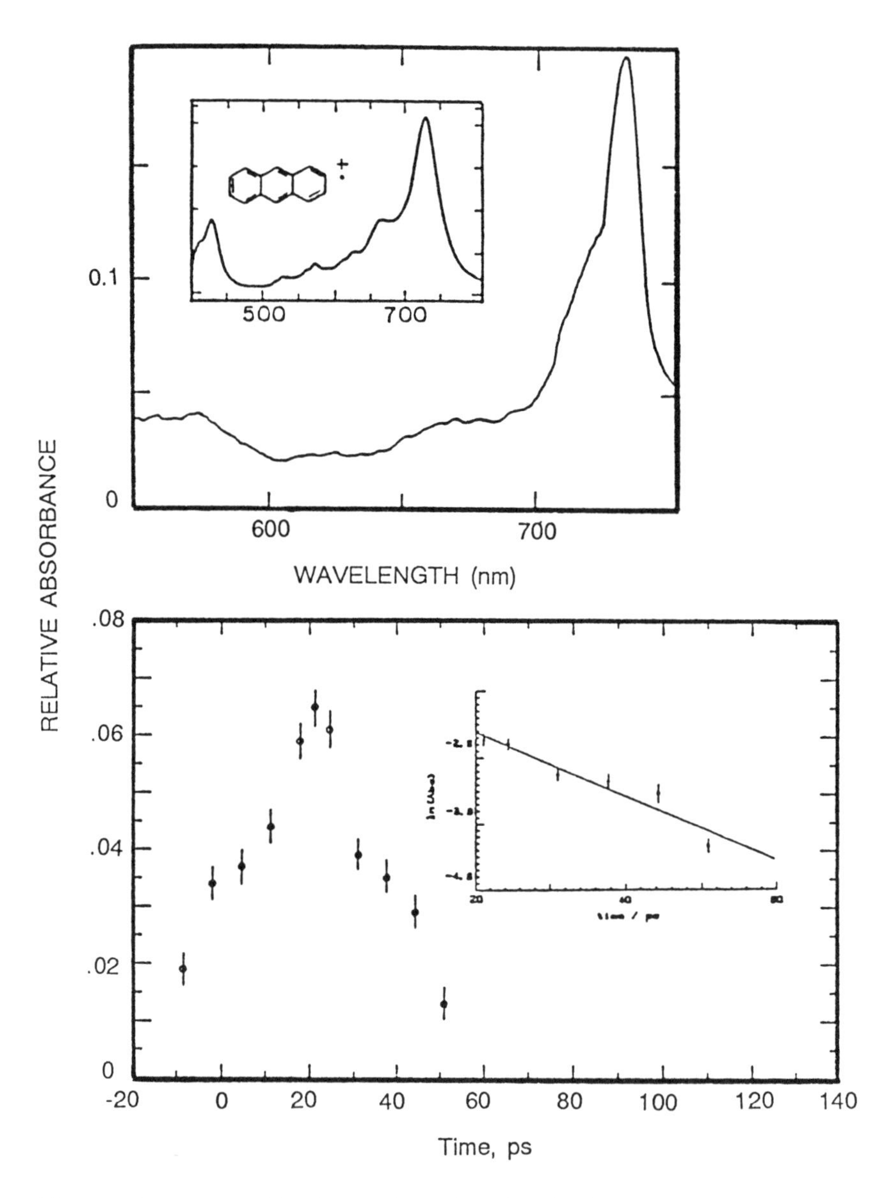

Figure 21. Top: Typical Time-Resolved Picosecond Absorption Spectrum Following the Charge-Transfer Excitation of Osmium Tetroxide EDA Complexes with Arenes (Anthracene) Showing the Growth of the Aromatic Cation Radical. Bottom: Temporal Evolution of $ArH^{\cdot+}$ Monitored at λ_{max}. Inset Shows the First-order Plot of the Ion Radical Decay.

thermal (PT), and charge-transfer (CT) osmylation. For example, the anthracenes react rather slowly with osmium tetroxide via the EDA complex to effect DT osmylation in nonpolar solvents and afford 2:1 adducts that are then converted to the more tractable pyridine derivatives such as **A**. Alternatively, the same ternary product **A** is directly formed at a significantly *enhanced* rate by the PT osmylation of anthracene with a mixture of OsO_4 and pyridine. Finally, the OsO_4 adduct to anthracene is instantly produced by CT osmylation involving photoexcitation of the [Ar,OsO_4] precursor complex. As such, the three procedures represent different activation mechanisms for arene oxidation. Thus DT and PT osmylations are adiabatic processes in which the transition states are attained via the collapse of an arene donor with the OsO_4 and the base-coordinated OsO_4 (py) electrophile, respectively. On the other hand, CT osmylation is a nonadiabatic process resulting from the vertical excitation of the [Ar, OsO_4] complex. For the latter, time-resolved picosecond spectroscopy can define the relevant photophysical and photochemical events associated with the charge-transfer excitation of an arene EDA complex, as previously established with arene complexes involving other electron acceptors. Accordingly, the CT osmylation is delineated first and then related to DT and PT osmylation. Before proceeding, however, it is important to emphasize that the DT, PT, and CT osmylations all share in common the formation of the 1:1 osmium(VI) cycloadduct $ArOsO_4$ in the initial rate-limiting step, since the concomitant loss of aromaticity produces a reactive alicyclic diene that is highly susceptible to the further thermal osmylation.[99] The universal adherence to the 2:1 adduct $Ar(OsO_4)_2$ (except phenanthrene), irrespective of the molar ratios of arene/OsO_4 and the particular procedure employed, accords with the rapid addition of a second mole of OsO_4 in DT, PT, and CT osmylations. This allows the focus on the formation of a single intermediate $ArOsO_4$ to delineate the unifying activation processes for DT, PT, and CT osmylations.

8.6. Electron Transfer in the Charge-Transfer Osmylation of Arenes

The direct observation of the reactive intermediates by the use of time-resolved picosecond spectroscopy and fast kinetics (Figure 21) enables the course of CT osmylation to be charted in some detail. The analysis proceeds from the mechanistic context involving the evolution and metamorphosis of the CT ion pair, as summarized in Scheme 14 the

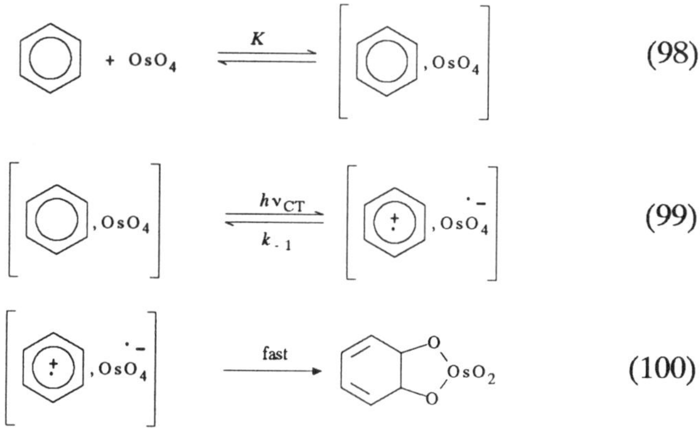

$$\text{C}_6\text{H}_6 + \text{OsO}_4 \underset{}{\overset{K}{\rightleftharpoons}} [\text{C}_6\text{H}_6, \text{OsO}_4] \quad (98)$$

$$[\text{C}_6\text{H}_6, \text{OsO}_4] \underset{k_{-1}}{\overset{h\nu_{CT}}{\rightleftharpoons}} [\text{C}_6\text{H}_6^{\cdot +}, \text{OsO}_4^{\cdot -}] \quad (99)$$

$$[\text{C}_6\text{H}_6^{\cdot +}, \text{OsO}_4^{\cdot -}] \xrightarrow{\text{fast}} \text{C}_6\text{H}_6\text{O}_2\text{OsO}_2 \quad (100)$$

Scheme 14.

critical initial step (Eq. 100) to form the 1:1 adduct to a benzene donor, where the brackets denote solvent-caged pairs.

All the experimental observations on CT osmylation indeed coincide with the formulation in Scheme 14. Thus the exposure of arene to osmium tetroxide leads immediately to new absorption bands (Figure 20) that are readily associated with the formation of the EDA complex in Eq. 98. These binary complexes are always present in low steady-state concentrations because of the limited magnitudes of K determined by the Benesi–Hildebrand method. The complexes are so weak that every attempt at isolation, including the freezing of various mixtures of OsO_4 in neat aromatic donors, merely leads to phase separation. The absorption bands are thus properly ascribed to contact charge transfer, as formulated by Orgel and Mulliken,[106] who predicted the CT absorption bands in these EDA complexes to be associated with the electronic excitation to the ion pair state (Eq. 99). As such, the time-resolved spectrum in Figure 20A identifies the formation of the arene cation radical to occur within the rise time of the 30-ps laser pulse. [The accompanying presence of the perosmate(VII) (OsO_4^-) counter-anion is obscured by the arene absorptions.] The electron transfer from the arene donor to the OsO_4 acceptor in the EDA complex (Eq. 99) effectively occurs with the absorption of the excitation photon ($h\nu_{CT}$), in accord with Mulliken's theory. Furthermore the appearance at < 30 ps demands that $Ar^{\cdot+}$ and $OsO_4^{\cdot-}$ are born as a contact (inner-sphere) ion pair with a mean separation essentially that of the precursor complex $[Ar, OsO_4]$ since this time scale obviates significant competition from diffusional processes. The seminal role of the ion pair $[Ar^{\bullet+}, OsO_4^{\bullet-}]$ as the obligatory intermediate from the photoexcitation of the EDA complex, must be included in any formulation of CT osmylation, by taking particular cognizance of how it decays. The spontaneous collapse of the CT ion pair in Eq. 100 represents the most direct pathway to arene cycloaddition—the measured half-life of $\tau \cong 35$ ps for the disappearance of the anthracene cation radicals in Figure 21B largely precluding diffusive separation of such ion pairs. However, the magnitudes of the product quantum yield $\Phi_P \sim 10^{-2}$ indicate that the primary route for ion pair decay is the back electron transfer as the reverse step of Eq. 99. An energy-wasting process with an estimated rate constant of $\sim 10^{11}$ s^{-1} derives from a highly exergonic driving force that is estimated to be $\mathcal{F}\Delta E^0 \cong -30$ kcal mol^{-1} based solely on the standard redox potentials of $E^0 = +1.30$ and -0.06 V for anthracene and the perosmate(VII) anion, respectively. More relevant to this issue is an estimated first-order rate constant for cycloaddition of $k_c = 10^9$ s^{-1} for the ion pair collapse to the arene cycloadduct in Scheme 14. Such a relatively large rate constant also points to a highly exergonic (bond-making) process for the cycloaddition in Eq. 100. Therefore the selectivity in adduct formation can be considered for various polynuclear arenes in which the initial addition of OsO_4 is possible at several sites. The regiospecificity observed in the CT osmylation of phenanthrene and 1,4-dimethylnaphthalene to produce only one isomeric adduct **P** and **N**

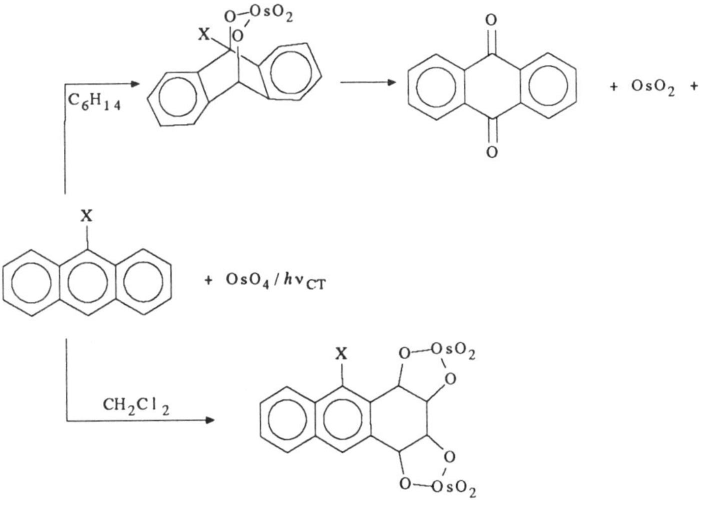

C_6H_{14} → [9-X-anthracene 9,10-O_2OsO_2 adduct] → [anthraquinone] $+ OsO_2 + HX$ (101)

[9-X-anthracene] $+ OsO_4/h\nu_{CT}$

CH_2Cl_2 → [bis-O_2OsO_2 adduct] (102)

respectively, accords with the reactive site centered on the arene HOMO. However, in the extended polynuclear anthracenes, the separation of the HOMO and subjacent SHOMO (i.e., HOMO-1) is not so well delineated, and the regiospecificity is strikingly modulated by solvent polarity. Ion pair annihilation is known to occur with the greatest ease in highly nonpolar alkanes.[74] Accordingly in *n*-hexane as solvent, the immediate collapse of the first-formed inner-sphere ion pair centered at the anthracene HOMO is expected to occur at the *meso* (9,10-) positions. Such an ion pair collapse would produce anthraquinone in a manner similar to that presented in Eq. 97. On the other hand, the formation of *only* adduct **A** from the initial addition of OsO_4 to the terminal ring represents a very unusual regiospecificity insofar as other addition (and substitution) reactions of anthracene are concerned. It suggests that the initially formed HOMO ion pair (HIP) has time to relax in the more polar dichloromethane medium to the isomeric SHOMO ion pair (SIP) that rapidly leads to adduct **A**. This proposal receives support from the observation of adducts related to **A** from the CT osmylation of both 9-methyl- and 9,10-dimethylanthracene in *hexane*. The enhanced stability of the cation radicals from these relatively electron-rich anthracenes will optimize the opportunity to convert the HIP to the more reactive SIP even in the nonpolar hexane medium, particularly if the collapse of the former were reversible.

relax (103)

(hexane) (dichloromethane)

8.7. Electron Transfer as the Common Theme in Arene Osmylation

The wide-ranging reactivity of various aromatic hydrocarbons to OsO_4 offers the unique opportunity to probe the activation process for oxidative osmylation, especially with regard to the role of the EDA complex and the reactive intermediates. In particular, the deliberate photoexcitation ($h\nu_{CT}$) of the EDA complex in hexane or dichloromethane effectively activates various arenes including benzenes, naphthalenes, and anthracenes to CT osmylation. This photoactivated process is readily associated with the charge-transfer ion pair,

$$[\mathrm{Ar}, \mathrm{OsO_4}] \xrightarrow{h\nu_{CT}} [\mathrm{Ar^{\bullet+}}, \mathrm{OsO_4^{\bullet-}}] \xrightarrow{\text{fast}} \mathrm{ArOsO_4}, \text{ etc.} \quad (104)$$

as established by the growth and decay of arene cation radicals with the aid of time-resolved picosecond spectroscopy. When kept in the dark, the same solutions of the EDA complexes slowly afford arene–OsO_4 adducts that are identical to those derived by CT osmylation. Indeed the close kinship between the thermal and charge-transfer activation of osmylation is underscored by the unique adduct **A** in which OsO_4 addition occurs exclusively to the terminal ring and not to the usual *meso* (9,10-) positions of anthracene. The activation process to form the kindred adiabatic ion pair [$Ar^{\bullet+}$, $OsO_4^{\bullet-}$] in the thermal osmylation provides the unifying theme in arene oxidation. Furthermore the promoted thermal osmylation of arenes via the 5-coordinate pyridine analogue OsO_4py is related to the widely used procedure for alkene bis-hydroxylation and the same regiochemistry observed, especially with anthracene donors indicates that the activated complex for PT osmylation is strongly related to that for DT osmylation.

The variable regiochemistry observed in the collapse of [$Ar^{\bullet+}$, $OsO_4^{\bullet-}$] to the cycloadduct $ArOsO_4$ underscores the importance of the contact (inner-sphere) ion pair structure in determining the course of electron-transfer oxidation. Since such structures are not readily determined as yet, the structural effects induced by qualitative changes in solvent polarity, salts, additives, and temperature are reaction variables that must always be optimized in the synthetic utilization of electron-transfer oxidation by either thermal or photochemical activation.

9. EPILOGUE

Aromatic hydrocarbons (ArH) are excellent electron donors by virtue of the ease of oxidation, as measured by the limited magnitudes of their 1-electron electrode potentials E^0_{Ar} in solution. The parallel values of the ionization potentials *IP* of the same series of arenes in the gas phase indicate that solvation energies are relatively insensitive to structural changes.

Electron transfer from various arenes to the one-electron acceptor tris-phenanthrolineiron(III) proceeds by an outer-sphere mechanism in which the second-order rate constant (log k_1) follows the driving force ($E^0_{Ar} - E^0_{Fe}$) in accord with Marcus theory. Arene cation radicals ($ArH^{\bullet+}$) formed by such an electron detachment from the HOMO of the aromatic donor are the critical intermediates whose thermodynamic stabilities, experimentally evaluated by E^0_{Ar}, relate to their lifetimes in solution—those cation radicals derived from the electron-rich dimethoxybenzene and methoxytoluene being the longest lived in comparison with the transient benzene cation radical.

Methylarene cation radicals show a marked increased kinetic acidity relative to their neutral precursor, as determined by both the direct and the indirect kinetics measurements of the second-order rate constants k_H for proton transfer to different pyridine bases. Such a deprotonation from a carbon center leads to the *sidechain* substitution of methylarenes. Competition from homolytic *nuclear* substitution (via the nucleophilic addition to $ArH^{\cdot+}$) can be readily evaluated by the magnitudes of the measured kinetic isotope effect for ring substitution ($k_H/k_D \cong 6$) in comparison with that observed for sidechain substitution ($k_H/k_D \cong 3$) of typical methylarene donors such as durene and methoxytoluene, as modulated by the steric hindrance in the pyridine base.

Alternatively, electron transfer is efficiently promoted by the actinic irradiation of the charge-transfer absorption bands of the electron donor–acceptor (EDA) precursor complexes of electrophiles with arenes. Such a photoinduced electron transfer is most useful for arenes with values of E^0_{Ar} that are too high (positive) to provide the driving force sufficient to effect the thermal (adiabatic) process, especially when used in conjunction with acceptors of limited strength such as tetranitromethane. For example, the latter forms a series of 1:1 EDA complexes with arenes [ArH, $C(NO_2)_4$], which upon charge-transfer excitation yields the reactive triad of cation radical, anion, and radical [$ArH^{\cdot+}$, $C(NO_2)_3^-$, $NO_2^{\cdot}$]. The collapse of the ion pair combination [$ArH^{\cdot+}$, $C(NO_2)_3^-$] leads to homolytic nuclear substitution (vide supra), whereas the collapse of the ion radical pair [$ArH^{\cdot+}$, $NO_2^{\cdot}$] leads to the Wheland intermediate pertinent to electrophilic aromatic substitution. The dichotomy between nuclear substitutions arising from the mutual annihilation of ion pairs and ion radical pairs is modulated by solvent polarity and "special" salt effects. Since the nuclear nitration of arene donors via the ion radical pair leads to isomer distributions that are indistinguishable from those obtained conventionally with nitric acid, it raises the more general question as to whether electrophilic aromatic substitutions proceed via prior electron transfer. Indeed the close parallel between charge-transfer and thermal processes is underscored in the addition of the electrophile osmium tetroxide to arenes. The characteristic solvent effects observed in the unique addition of OsO_4 to the terminal ring of anthracene provides the unifying mechanistic theme for aromatic activation via electron transfer.

ACKNOWLEDGMENTS

We thank our co-workers mentioned in the references and especially C. J. Schlesener, J. M. Masnovi, S. Sankararaman, and J. M. Wallis for their creative and dedicated efforts. C. A. thanks the Centre National de la Recherche Scientifique and J. K. K. thanks the National Science Foundation, R. A. Welch Foundation, and the Texas Advanced Research Program for financial support. We also thank the U. S.–France Cooperative Research Program jointly sponsored by NSF and CNRS for travel funds.

REFERENCES

1. (a) Andrulis, P. J., Jr.; Dewar, M. J. S.; Dietz, R.; Hunt, R. L. *J. Am. Chem. Soc.* **1966**, *88*, 5473. (b) Aratani, T.; Dewar, M. J. S. *J. Am. Chem. Soc.* **1966**, *88*, 5479. (c) Andrulis, P. J., Jr.; Dewar, M. J. S. *J. Am. Chem. Soc.* **1966**, *88*, 5483.
2. Kochi, J. K.; Tang, R. T.; Bernath, T. *J. Am. Chem. Soc.* **1973**, *95*, 7114.
3. For reviews, see (a) Kochi, J. K. In *Free Radicals.* Wiley, New York, 1973, Vol. 1, Chapter 11. (b) Hammerich, O.; Parker, V. D. *Adv. Phys. Org. Chem.* **1984**, *20*, 55.
4. Eberson, L. *Electron Transfer in Organic Chemistry.* Springer-Verlag, New York, 1987.
5. Yoshida, K. *Electrooxidation in Organic Chemistry.* Wiley, New York, 1984.
6. Kochi, J. K. *Angew. Chem. Int. Ed. Eng.* **1988**, *27*, 1227.
7. Klingler, R. J.; Kochi, J. K. *J. Am. Chem. Soc.* **1982**, *104*, 4186.
8. For a review, see Schofield, K. *Aromatic Nitration.* Cambridge University Press, Cambridge, 1980.
9. (a) Kenner, J. *Nature (London)* **1945**, *156*, 369. (b) Pederson, E. B.; Petersen, T. E.; Torsell, K.; Lawesson, S. O. *Tetrahedron* **1973**, *29*, 579. (c) Perrin, C. L. *J. Am. Chem. Soc.* **1977**, *99*, 5516.
10. Lewis, E. S. In *Investigation of Rates and Mechanisms of Reactions*, Weissberger, A., Ed., Part I. Wiley, New York, 1961, Chapter 1.
11. Kochi, J. K. *Acta Chem. Scand.*, **1990**, *44*, 409–43.
12. (a) Mijs, W. J.; deJonge, C. R. H. I., Eds. *Organic Synthesis by Oxidation with Metal Compounds.* Plenum, New York, 1986. (b) For a recent review see Kochi, J. K. In *Comprehensive Organic Synthesis*, Ley, S. V., Ed. Pergamon, New York, Vol. 7, in press.
13. Meites, L.; Zuman, P. *CRC Handbook Series in Organic Electrochemistry.* CRC Press, Cleveland, 1981.
14. See Wightman, R. M. *Anal. Chem.* **1981**, *53*, 1125A.
15. Howell, J. O.; Goncalves, J. M.; Amatore, C.; Klasinc, L.; Wightman, R. M.; Kochi, J. K. *J. Am. Chem. Soc.* **1984**, *106*, 3968.

16. Fritz, H. P.; Artes, R. O. *Electrochim. Acta* **1981**, *26*, 417; Hammerich, O,; Parker, V. D. *Electrochim. Acta* **1973**, *18*, 537. See also Kaiser, E. T.; Kevan, L. O., Eds. *Radical Ions*. Wiley, New York, 1968.
17. See also (a) Kimura, K.; Katsumata, S.; Achiba, Y.; Yamazaki, T.; Iwata, S. *Handbook of He(I) Photoelectron Spectra of Organic Molecules*. Halstead, New York, 1981. (b) Klasinc, L.; Kovac, B.; Gusten, H. *Pure Appl. Chem.* **1983**, *55*, 289.
18. (a) Amatore, C.; Jutand, A.; Pflüger, F. *J. Electroanal. Chem.* **1987**, *218*, 361. (b) Amatore, C.; Lefrou, C.; Pflüger, F. *J. Electroanal. Chem.* **1989**, *207*, 43. (c) Amatore, C.; Lefrou, C. Unpublished results, **1989**. (d) Potentials are expressed in volts vs SCE. $E^0_{ferrocene}$ was measured as 0.371 V with the same reference.[19]
19. Compare Gagne, R. R.; Koval; C. A.; Lisensky, G. C. *Inorg. Chem.* **1980**, *19*, 2854.
20. When both the anodic and cathodic CV waves are visible as in Figures 1 and 2, the standard oxidation potentials E^0 must lie in the interval $[E^c_p + 30 mV] < E^0 < [E^a_p - 30 mV]$. See Nicholson, R. S. *Anal. Chem.* **1965**, *37*, 1351.
21. The effect is largely due to contributions from $ArH^{\bullet+}$ since the solvation energies of uncharged species are small. See Abraham, M. H. *J. Am. Chem. Soc.* **1982**, *104*, 2085 and Lofti, M.; Roberts, R. M. G. *Tetrahedron* **1979**, *35*, 2137.
22. (a) Intermediate case is represented by 1,2,3,4-tetramethylbenzene (TMB). (b) Based on E^0_{Ar} in volts vs $Ag/AgClO_4$ and *IP* in electron volts.
23. See, e.g., Bockris, J. O'M.; Reddy, A. K. N. *Modern Electrochemistry*. Plenum, New York, 1970, Vol. I, 56ff. $\Delta G^0_s = (Ne^2/2r)(1-1/D)$ where r is the radius of the equivalent sphere to the ion and D = 39.5 is the dielectric constant for trifluoroacetic acid.
24. For example, ΔG^0_s = 23.5 and 20.9 kcal mol^{-1} for toluene and hexamethylbenzene with r = 3.48 and 3.93 Å, respectively. The difference thus accounts for more than half of the solvation energy difference between Eqs. 15 and 16.
25. Schlesener, C. J.; Amatore, C.; Kochi, J. K. *J. Am. Chem. Soc.* **1984**, *106*, 3567.
26. Rollick, K. L.; Kochi, J. K. *J. Am. Chem. Soc.* **1982**, *104*, 1319.
27. E^0_{Ar} = 1.58, 1.69, 1.75, and 1.77 V vs SCE in acetonitrile containing 0.6 *M* TBAB (based on E^0_{Fe} = 0.371 V for ferrocene in the same potential scale) and E^0_{Fe} = 1.09, 1.19, and 1.29 V for X = H, 5-Cl, and 5-NO_2 substituted phenanthrolineiron(III) complexes.
28. See, e.g., Moore, J. W.; Pearson, R. G. In *Kinetics and Mechanism*. Wiley Interscience, New York, 1981, and Cannon, R. D. In *Electron Transfer Reactions*. Butterworths, London, 1980.
29. Cf. Scandola, F.; Balzani, V.; Schuster, G. B. *J. Am. Chem. Soc.* **1981**, *103*, 2519. Andrieux, C. P.; Blocman, C.; Dumas-Bouchiat, J. M.; Saveant, J. M. *J. Am. Chem. Soc.* **1979**, *101*, 3431.
30. See, e.g., Marcus, R. A. *Faraday Discuss. Soc.* **1960**, *29*, 129 and Marcus, R. A.; Siders, P. In *Mechanistic Aspects of Inorganic Reactions*, Rorabacher, D. B.; Endicott, J. F., Eds. ACS Symposium Series, Washington, C.D., 1982, Chapter 10, pp. 235–238.
31. For the computations, (a) the diffusion rate constants k_r and k_p for Eq. 26 were taken to be 4×10^{10} and 6×10^9 M^{-1} s^{-1}, respectively. (b) The work term $w_r \cong 0$

and $w_P \cong 1.5$ kcal mol^{-1} was obtained from an electrostatic model of $ArCH_3^{\bullet+}$ and FeL_3^{2+} at a distance, $d^{\ddagger} = r_{Ar} + r_{Fe} = 3.5 + 7.0 = 10.5$ Å. Note from the plot in Figure 5, $w_P = 1 \pm 1$ kcal mol^{-1}.
32. Brunschwig, B. S.; Creutz, C.; Macartney, D. H.; Sham, T.-K.; Sutin, N. *Faraday Discuss. Chem. Soc.* **1982**, *74*, 113.
33. Bevington, P. R. In *Data Reduction and Error Analysis for the Physical Sciences.* McGraw-Hill, New York, 1969, p. 137.
34. See Klingler, R. J.; Kochi, J. K. in ref. 7 for highly endergonic reactions in which k_{-1} can approach the diffusion-controlled limit of 10^{10} M^{-1} s^{-1}.
35. (a) Makajima, T.; Toyota, A.; Kataoka, M. *J. Am. Chem. Soc.* **1982**, *104*, 5610 and references therein. (b) Iwasaki, M.; Toriyama, K.; Nunome, K. *J. Chem. Soc., Chem. Commun.* **1983**, 320.
36. Salem, L. *The Molecular Orbital Theory in Conjugated Systems.* Benjamin, New York, 1966, pp. 467–485.
37. Dickens, J. E.; Basolo, F.; Neumann, H. M. *J. Am. Chem. Soc.* **1957**, *79*, 1289.
38. We are actually faced with a conundrum here, since arguments such as those based on the Hammond postulate are most applicable at the endergonic and exergonic limits where the contribution from activation ($\Delta G^{\ddagger}$ or $\Delta H^{\ddagger}$) is minor, and the rate is controlled by diffusion. Furthermore most kinetic measurements are not carried out at the diffusion limits.
39. Sheldon, R. A.; Kochi, J. K. *Metal-Catalyzed Oxidations of Organic Compounds.* Academic Press, New York, 1981. See also refs. 3 and 5.
40. Landau, R.; Saffer, A. *Chem. Eng. Prog.* **1968**, *64*, 20.
41. Bewick, A.; Edwards, G. J.; Mellor, J. M.; Pons, B. S. *J. Chem. Soc. Perkin Trans. 2*, **1977**, 1952; Bewick, A.; Mellor, J. M.; Pons, B. S. *Electrochim. Acta* **1980**, *25*, 931.
42. Schlesener, C. J.; Amatore, C.; Kochi, J. K. *J. Am. Chem. Soc.* **1984**, *106*, 7472.
43. Z is taken as 10^{11} M^{-1} s^{-1} and $\kappa = 1$. Note that an experimental uncertainty of ±0.010 V in the determination of $(E^0_{Ar} - E^0_{Fe})$ leads to an uncertainty of ~±0.25 kcal mol^{-1} in $\Delta G^{\ddagger}$ in Table V.
44. Masnovi, J. M.; Sankararaman, S.; Kochi, J. K. *J. Am. Chem. Soc.* **1989**, *111*, 2263.
45. (a) Note many of the values of pK_a^B in Tables IV and V refer to the acid dissociation constants of the pyridinium ions determined directly in acetonitrile solution by Cauquis, G.; Deronzier, A.; Serve, D.; Vieil, E. *Electroanal. Chem. Interfac. Electrochem.* **1975**, *60*, 205. (b) Others were obtained from a correlation of the pK_a^B value in water [Perrin, D. D.; Dempsey, B.; Serjeant, E. P. *pKa Prediction from Organic Acids and Bases.* Chapman & Hall, London, 1981.] with that in acetonitrile (vide supra). (c) The value for 2,6-di-*t*-butyl pyridine was from Brown, H. C.; Kanner, B. *J. Am. Chem. Soc.* **1966**, *88*, 986 determined with others in 50% aqueous ethanol, and correlated against those in water.
46. (a) Marcus, R. A. *J. Phys. Chem.* **1968**, *72*, 891. (b) Cohen, A. O.; Marcus, R. A. *J. Phys. Chem.* **1968**, *72*, 4249. (c) Marcus, R. A. *J. Am. Chem. Soc.* **1969**, *91*, 7224. For the application of the Marcus equation to proton transfer, see (d) Kreevoy, M. M.; Konasewich, D. E. *Adv. Chem. Phys.* **1972**, *21*, 243. (e) Hupe, D. J.; Wu, D. *J. Am. Chem. Soc.* **1977**, *99*, 7653. (f) Kreevoy, M. M.; Oh, S.-W.

J. Am. Chem. Soc. **1973**, *95*, 4805. (g) Toullec, J. *Adv. Phys. Org. Chem.* **1982**, *18*, 5.

47. (a) Albery, W. J.; Kreevoy, M. M. *Adv. Phys. Org. Chem.* **1978**, *16*, 87. (b) Lewis, E. S. *J. Am. Chem. Soc.* **1989**, *111*, 7576 and references therein.
48. We make a distinction here between Marcus *theory* in the derivation of Eq. 31, which can be employed as the Marcus *equation* to describe the functional form of free energy relationships. Under the latter circumstance (including most inner--sphere processes), the physical significance of $\Delta G^{\ddagger}$, w_r, and w_p differs from those relating to outer-sphere electron transfer as originally conceived by Marcus.[49,50]
49. Marcus, R. J.; Zwolinski; B. J.; Eyring, H. *J. Phys. Chem.* **1954**, *58*, 432. Marcus, R. A. *J. Chem. Phys.* **1956**, *24*, 4966; *J. Chem. Phys.* **1957**, *26*, 867; *J. Chem. Phys.* **1965**, *43*, 679; *Discuss. Faraday Soc.* **1960**, *29*, 21.
50. For reviews, see (a) Sutin, N. In *Inorganic Biochemistry*, Eichhorn, G. L. Ed. Elsevier, Amsterdam, 1973, Vol. 2, p. 611. (b) Reynold, W. L.; Lumry, R. W. *Mechanism of Electron Transfer*. Ronald Press, New York, 1965. See also Cannon, R. D. in ref. 28.
51. A statistical correction for the number of available protons is not included[52] to also affect the values of w_p and w_r.
52. (a) Bell, R. P. *The Proton in Chemistry*, 2nd ed. Cornell University Press, Ithaca, NY, 1973. (b) Caldin, E.; Gold, V. Eds. *Proton Transfer Reactions*. Chapman & Hall, London, 1974.
53. Arene cation radicals $ArH^{\bullet+}$ are known to form π-complexes with arenes in the form of dimer cation radicals $[ArH]_2^{\bullet+}$ [See Edlund, O.; Kinell, P.-O; Lund, A.; Shimizu, A. *J. Chem. Phys.* **1967**, *46*, 3679. Badger, B.; Brocklehurst, B. *Trans. Faraday Soc.* **1969**, *65*, 2582.] For the structural effects of donor–acceptor complexes between aromatic moieties, see R. Foster, *Organic Charge Transfer Complexes*. Academic Press, New York, 1969.
54. The precursor complex(es) may also be viewed in other structural forms, including those with linear C···H···N interactions.
55. Nicholas, A. M. P.; Arnold, D. R. *Can. J. Chem.* **1982**, *60*, 2165.
56. The values of pK_a^A for the radicals of PMB, DUR, and TMB are evaluated by this method as –2, –3, and –3, respectively.
57. (a) Beletskaya, I. P.; Makhon'kov, D. I.; *Russ. Chem. Rev.* **1981**, *50*, 534. (b) Tomilov, A. P. *Russ. Chem. Rev.* **1961**, *30*, 639. (c) Eberson, L.; Nyberg, K. *J. Am. Chem. Soc.* **1966**, *88*, 1686. (d) Eberson, L.; Nyberg, K. *Acc. Chem. Res.* **1973**, *106*, 6. (e) Komatsu, T.; Lund, A.; Kinell, P.-O. *J. Phys. Chem.* **1972**, *76*, 1721.
58. (a) Heiba, E. I.; Dessau, R. M.; Koehl, W. J., Jr. *J. Am. Chem. Soc.* **1969**, *41*, 6830. (b) Hanotier, J.; Hanotier-Bridoux, M.; de Radzitzky, P. J. *J. Chem. Soc. Perkin* **1973**, *2*, 381. (c) Ross, S. D.; Finkelstein, M.; Peterson, R. C. *J. Org. Chem.* **1970**, *35*, 781. (d) Blum, Z.; Cedheim, L.; Nyberg, K. *Acta Chem. Scand.* **1975**, *B29*, 715. (e) Shaw, M. J.; Weil, J. A.; Hyman, H. H.; Filler, R. *J. Am. Chem. Soc.* **1970**, *92*, 5096. (f) Shaw, M. J.; Hyman, H. H.; Filler, R. *J. Org. Chem.* **1971**, *36*, 2918. (g) Eberson, L.; Oberrauch, E. *Acta Chem. Scand.* **1981**, *B35*, 193. (h) Taube, H. *Angew. Chem. Int. Ed.* **1984**, *23*, 329.
59. Schlesener, C. J.; Kochi, J. K. *J. Org. Chem.* **1984**, *49*, 3142.
60. (a) Eberson, L. *J. Am. Chem. Soc.* **1983**, *105*, 3192. (b) Eberson, L. *J. Am. Chem. Soc.* **1967**, *89*, 4669. (c) Eberson, L.; Jonsson, L.; Wistrand, L.-G. *Acta Chem.*

Scand. **1978**, *B32*, 520. (d) Nyberg, K.; Wistrand, L.-G. *J. Org. Chem.* **1978**, *43*, 2613. (e) Marrocco, M.; Brilmyer, G. *J. Org. Chem.* **1983**, *48*, 1487. (f) Nilsson, A.; Palmquist, U.; Petterson, T.; Ronlan, A. *J. Chem. Soc. Perkin Trans. I*, **1978**, 708. (g) Uemura, S.; Ikeda, T.; Tanaka, S.; Okano, M. *J. Chem. Soc. Perkin I*, **1979**, 2574. See also Dewar et al. in ref. 1.

61. For the difficulty associated with accurate values[43] of (E^0_{Fe}–E^0_{Ar}) in the determination of k_2, see the discussion in section 4.3.
62. (a) Parker, V. D.; Eberson, L. *Tetrahedron Lett.* **1969**, 2843. (b) Parker, V. D.; Eberson, L. *Tetrahedron Lett.* **1969**, 2839. (c) Manning, G.; Parker, V. D.; Adams, R. N. *J. Am. Chem. Soc.* **1969**, *91*, 4584. (d) Lund, H. *Acta Chem. Scand.* **1957**, *11*, 1323.
63. Yoshida, K.; Shigi, M.; Fueno, T. *J. Org. Chem.* **1975**, *40*, 63. See also Andreades, S.; Zahnow, E. W. *J. Am. Chem. Soc.* **1969**, *91*, 4181.
64. See Bell, R. P. in ref. 52a.
65. See Cauquis, G. et al. in ref. 45a.
66. See Kenner, J. in ref. 9a.
67. Brown, R. D. *J. Chem. Soc.* **1959**, 2224, 2232.
68. (a) Nagakura, S.; Tanaka, J. *J. Chem. Phys.* **1954**, *22*, 563. (b) Nagakura, S. *Tetrahedron*, **1963**, *19* (Suppl. 2), 361.
69. Eberson, L.; Radner, F. *Acta Chem. Scand.* **1980**, *B34*, 739. See also Schofield, K. in ref. 8 and Pederson, E. B. et al. in ref. 9b. In addition see Perrin, C. L. ref. 9c.
70. Wheland, G. W. *J. Am. Chem. Soc.* **1942**, *64*, 900.
71. (a) Hartshorn, S. R. *Chem. Soc. Rev.* **1974**, *3*, 167. (b) Suzuki, H. *Synthesis* **1977**, 217.
72. (a) Masnovi, J. M.; Hilinski, E. F.; Rentzepis, P. M.; Kochi, J. K. *J. Am. Chem. Soc.* **1986**, *108*, 1126. (b) Masnovi; J. M.; Kochi, J. K. *J. Am. Chem. Soc.* **1985**, *107*, 7880. (c) Sankararaman, S.; Haney, W. A.; Kochi, J. K. *J. Am. Chem. Soc.* **1987**, *109*, 5235, 7824.
73. Mulliken, R. S.; *J. Am. Chem. Soc.* **1952**, *74*, 811. Mulliken, R. S.; Person, W. B. *Molecular Complexes.* Wiley, New York, 1969.
74. Masnovi, J. M.; Kochi, J. K. *J. Am. Chem. Soc.* **1985**, *107*, 6781. See also ref. 72b.
75. Melander, L. *Arkiv Kemi* **1950**, *2*, 211; Halvarson, K.; Melander, L. *Arkiv Kemi* **1957**, *11*, 77. See also Melander, L. *Isotope Effects on Reaction Rates.* Ronald Press, New York, 1960.
76. Fukuzumi, S.; Kochi, J. K. *J. Am. Chem. Soc.* **1981**, *103*, 7240.
77. Fischer, A.; Fyles, D. L.; Henderson, G. N. *J. Chem. Soc. Chem. Commun.* **1980**, 513. See also Fisher, A.; Leonard, D. R. A.; Roderer, R. *Can. J. Chem.* **1979**, *57*, 5727.
78. Clark, E. P. *J. Am. Chem. Soc.* **1931**, *53*, 3431.
79. Loudon, J. D.; Ogg, J. *J. Chem. Soc.* **1955**, 739.
80. (a) Perrin, C. L.; Skinner, G. A. *J. Am. Chem. Soc.* **1971**, *93*, 3389. (b) Bunton, C. A.; Hughes, E. O.; Ingold, C. K.; Jacobs, D. J. H.; Jones, M. H.; Minkoff, G. J.; Reed, R. I. *J. Chem. Soc.* **1950**, 2628. (c) Reverdin, F.; During, F. *Chem. Ber.* **1899**, *32*, 152. (d) Robinson, G. M. *J. Chem. Soc.* **1916**, *109*, 1078. (e) Barnes, C. E.; Feldman, K. S.; Johnson, M. W.; Lee, H. W. H.; Myhre, P. C. *J. Org. Chem.* **1979**, *44*, 3925. (f) Barnes, C. E.; Myhre, P. C. *J. Am. Chem. Soc.* **1978**, *100*, 973.
81. (a) Bandlish, B. K.; Shine, J. J. *J. Org. Chem.* **1977**, *42*, 561, (b) Williams, D. L. H. *Nitrosation.* Cambridge University Press, Cambridge, 1988.

82. Brownstein, S.; Gabe, E.; Lee, F.; Piotrowski, A. *Can. J. Chem.* **1986**, *64*, 1661.
83. Kim, E. K.; Kochi, J. K. *J. Org. Chem.* **1989**, *54*, 1692.
84. Kampar, E.; Neilands, O. *Russ. Chem. Rev. SS* **1986**, 334.
85. See (a) Kampar, V. E.; Valtere, S. P.; Neiland, O. Ya. *Theor. Exp. Chem. SSR* **1978**, *14*, 288. (b) Kampar, V. E.; Gudele, I. Ya.; Neiland, O. Ya. *Theor. Exp. Chem. SSR* **1980**, *16*, 321. (c) Chappell, J. S.; Block, A. N.; Bryden, W. A.; Maxfield, M.; Poehler, T. O.; Cowan, D. O. *J. Am. Chem. Soc.* **1981**, *103*, 2442.
86. Hanna, M. W. In Foster, R. Ed. *Molecular Complexes*. Elek Science, London, 1973, p. 1ff.
87. Sankararaman, S. unpublished results.
88. See Kaiser, E. T.; Kevan, L. O. in ref. 16.
89. Perkins, M. J. In *Free Radicals*, Kochi, J. K. Ed. Wiley, New York, 1973, Vol. 2, p. 231ff.
90. Masnovi, J. M.; Seddon, E. A.; Kochi, J. K. *Can. J. Chem.* **1984**, *62*, 2552.
91. (a) Fukuzumi, S.; Kochi, J. K. *J. Phys. Chem.* **1980**, *84*, 608 and related papers. (b) Blackstock, S. C.; Kochi, J. K. *J. Am. Chem. Soc.* **1987**, *109*, 2484.
92. Koenig, T.; Fischer, H. in *Free Radicals*, Kochi, J. K. Ed. Wiley, New York, 1973, Vol. 1, p. 157ff.
93. Nonhebel, D. C.; Walton, J. C. *Free Radical Chemistry*. Cambridge University Press, London, 1974, p. 447ff.
94. (a) Winstein, S.; Klinedinst, P. E., Jr.; Robinson, G. C. *J. Am. Chem. Soc.* **1961**, *83*, 885. (b) Winstein, S.; Klinedinst, P. E., Jr.; Clippinger, E. *J. Am. Chem. Soc.* **1961**, *83*, 4986. (c) Winstein, S.; Robinson, G. C. *J. Am. Chem. Soc.* **1958**, *80*, 169 and related papers. For reviews see (d) Harris, J. M. *Prog. Phys. Org. Chem.* **1974**, *11*, 89. (e) Szwarc, M. Ed. *Ions and Ion Pairs in Organic Reactions*. Wiley-Interscience, New York, 1974, Vol. 2, Chapter 3, p. 247ff.
95. Goodson, B. E.; Schuster, G. B. *J. Am. Chem. Soc.* **1984**, *106*, 7254.
96. Hammond, G. S. *J. Am. Chem. Soc.* **1955**, *77*, 334.
97. See Fukuzumi, S. in ref. 76.
98. (a) Holm, R. H. *Chem. Rev.* **1987**, *87*, 1401. (b) Jorgensen, K. A. *Chem. Rev.* **1989**, *89*, 431.
99. (a) Hofmann, K. A. *Chem. Ber.* **1912**, *45*, 3329. (b) Hofmann, K. A.; Ehrhart, O.; Schneider, O. *Chem. Ber.* **1913**, *46*, 1657. (c) Criegee, R. *Liebigs. Ann.* **1936**, *522*, 75; (d) Criegee, R.; Marchand, B.; Wannowius, H. *Liebigs. Ann.* **1942**, *550*, 99. (e) Schroder, M. *Chem. Rev.* **1980**, *80*, 187.
100. (a) Nugent, W. A. *J. Org. Chem.* **1980**, *45*, 4533. (b) Hammond, P. R.; Lake, R. R. *J. Chem. Soc. (A)* **1971**, 3819. (c) Burkhardt, L. A.; Hammond, P. R.; Knipe, R. H.; Lake, R. R. *J. Chem. Soc. (A)* **1971**, 3789.
101. (a) Cook, J. W.; Schoental, R. *J. Chem. Soc.* **1948**, 170. (b) Cook, J. W.; Loudon, J. D.; Williamson, W. F. *J. Chem. Soc.* **1950**, 911. (c) Criegee, R.; Hoger, E.; Huber, G.; Knick, P.; Martscheffel, F.; Schellenberger, H. *Liebigs Ann.* **1956**, *599*, 81.
102. Cook, J. W.; Schoental, R. *Nature (London)* **1948**, *161*, 237.
103. (a) Hilinski, E. F.; Masnovi, J. M.; Kochi, J. K.; Rentzepis, P. M. *J. Am. Chem. Soc.* **1984**, *106*, 8071. (b) See also ref. 72.
104. Wallis, J. M.; Kochi, J. K. *J. Am. Chem. Soc.* **1988**, *110*, 8207.
105. Takahashi, Y.; Sankararaman, S.; Kochi, J. K. *J. Am. Chem. Soc.* **1989**, *111*, 2954.
106. Orgel, L. E.; Mulliken, R. S. *J. Am. Chem. Soc.* **1957**, *79*, 4839.

DISTANCE AND ANGLE EFFECTS ON ELECTRON TRANSFER RATES IN CHEMISTRY AND BIOLOGY

George L. McLendon and Anna Helms

1. Introduction 149
2. Effects of Donor–Acceptor Angle on Electron Transfer 153
3. Applications in Biological Electron Transfer 157
    3.1. The Photosynthetic Reaction Center 157
    3.2. Other Biological Systems 158
References 161

1. INTRODUCTION

Other chapters in monographs in this series will detail the role of electron transfer reactions in a variety of chemical processes and outline the basic theory of this class of reactions. In this chapter, we will focus on the role of wavefunction overlap, as modified by the donor–acceptor distance and angle, in governing electron transfer reactions with particular examples from biology.

Advances in Electron Transfer Chemistry,
Volume 1, pages 149–161.

ISBN: 1-55938-167-1

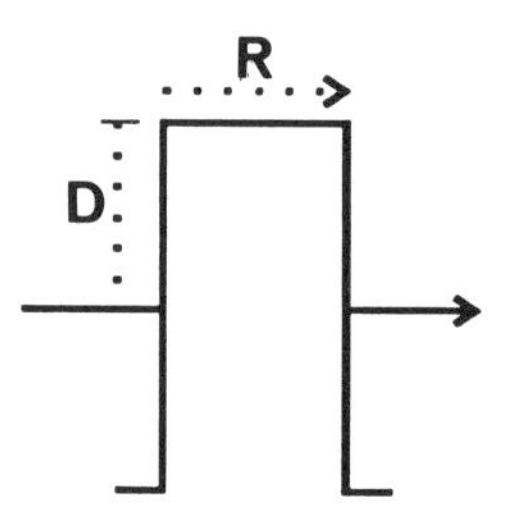

Figure 1. Probability of Electron Tunneling Through a Classically "Impenetrable" Barrier (After Gamow[3a]).

The strong distance dependence of electron transfer is governed by the radial component of the reactant wavefunctions; the probability of finding a valence electron far away (>VDW radius) from a molecule is small. However, it is well known that electron transfer reactions can occur at respectable rates even when the reactants are separated by a distance far greater than van der Waals (VDW) contact.

For example, pioneering studies of electron transfer in rigid glasses by Miller[1,2] gave evidence that electron transfer could occur at a rate of ~7 s^{-1} even with a 30 Å donor acceptor separation.[2] As the donor and acceptor approach more closely, the rate constant increased roughly exponentially $k = C(\exp(-\alpha R))$, $\alpha \sim 1.2$ Å^{-1}; that is, the rate constant changes 10-fold for every 2 Å change in distance (C is a constant).

Several theoretical explanations qualitatively predict this strong dependence of rate on reactant separation.[3] The oldest and simplest explanation is derived from basic quantum mechanics (Gamow Tunneling).[3a] The electron is considered to be confined to either the donor or acceptor separated by a "classically impenetrable" barrier made up of the intervening medium (Figure 1). The barrier height is set by the difference (D) in ionization potential between the electron donor and the medium. According to classical physics, the transfer can occur only over the barrier, so the (maximum) rate constant for electron transfer in this two state system should be $k_{et} \sim (kT/\hbar)$ ($\exp -D/kT$). For $D \sim 2$ eV (a reasonable value), the maximum rate would be 10^{26} s, which is longer than the age of the universe! Furthermore, the rate would be independent of the reactant separation (barrier width), R, in contrast to the strong dependence observed. In quantum physics, the situation is quite different.

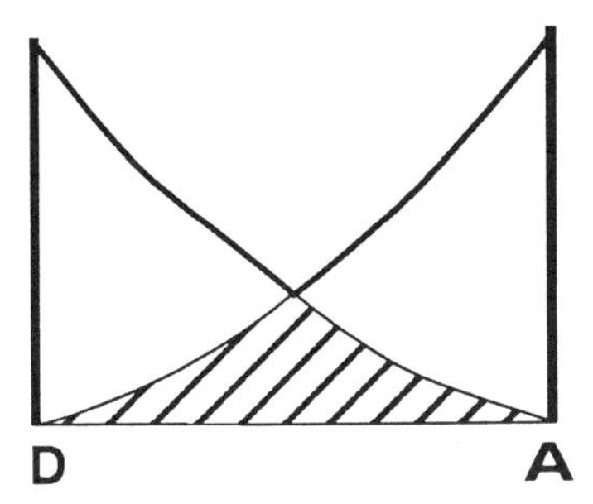

Figure 2. Schematic of Wavefunction Overlap Between a Donor and Acceptor Held Fixed at a Distance >> van der Waals Radius.

The uncertainty principle shows that the position and energy of a particle cannot simultaneously be *exactly* defined. Thus, if the reactant separation is small, then, by defining *D exactly*, the uncertainty in position is sufficiently large that there is some probability that the electron will already be *at* the acceptor. This probability necessarily decreases strongly as the uncertainty in distance decreases (i.e., as distance increases). Gamow showed this probability of barrier tunneling scales as P = exp-$(CDm)^{1/2}R$. The strong dependence on mass suggests that tunneling is quite common. Indeed, in all electron transfer reactions, at the instant of transfer, the electron always "tunnels," though R may be quite small. Conversely, the probability of tunneling becomes negligible as M increases: proton tunneling is common, but diatomic tunneling is rare (but known), and it is well known, to those of us who have tried, that people cannot walk through walls.

Although this barrier tunneling model has the elegance of simplicity, it clearly lacks chemical detail. Therefore, two more explicitly "chemical" models, which also predict the observed exponential dependence of rate on distance, will be mentioned. The first is based on direct donor–acceptor orbital overlap.[3b] Recall that hydrogen (like) wavefunctions decay exponentially beyond the Bohr radius. Thus, for two wavefunctions (donor and acceptor), the probability of wavefunction overlap should scale as exp-(BR) (Figure 2). For two hydrogen atoms separated by vacuum, $B = 2$, but for any realistic situation, the precise value of the exponential damping factor will depend not only on the molecule being studied, but also on the way in which the electron wavefunction of the "tunneling" electron couples to the intervening medium. This coupling

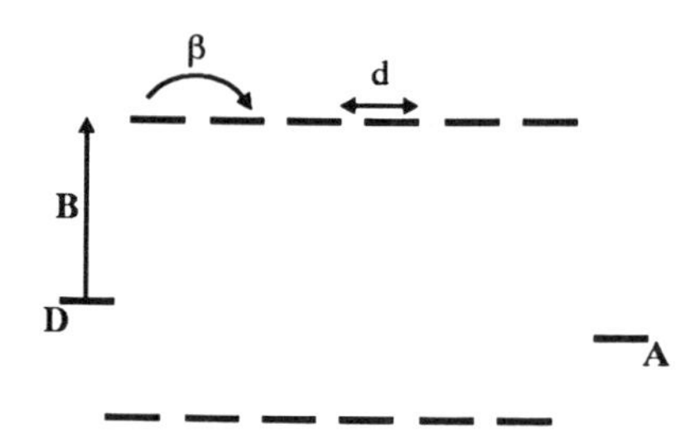

Figure 3. Schematic Diagram for Donor–Acceptor Superexchange Coupling.

is treated explicitly in the "superexchange model" of Beitz and Miller,[3d] developed to explain unusually fast rates of long distance electron transfer involving strong oxidants (hole transfer). This model is based in its essence on earlier work by McConnell.

In this model, the electron on a donor (or the "electron hole" on an acceptor) couples to the intervening medium (e.g., solvent) by a charge transfer interaction. The strength of this interaction (the coupling constant) depends critically on the energy mismatch between the donor and medium, as in the Gamow tunneling model already discussed. However, the overall matrix element, which governs transfer probability, depends not only on this energy match but also on propagation through the medium. It is assumed that electronic coupling propagates via an exchange interaction (directly analogous to the Hückel "β" exchange integral) (Figure 3). The *overall* probability for transfer can then be formulated to give the observed exponential dependence of rate on distance $k = C(\exp(-\alpha R))$ with $\alpha \equiv (1/d)\ \ln(B/\beta)$.

We note, however, that this function predicts quite a different dependence of rate on the classical energy deficit (D in the Gamow equation). Although the barrier tunneling model demands $k \propto (D)^{1/2}$, the superexchange model predicts a "kinder, gentler" dependence: $k \propto \ln(D)^{1/2}$. Current experimental data offer some support for the superexchange model. Specifically, when α is varied over a wide range, k depends weakly on D; $k \approx (\text{constant})(\ln D)$. For a useful comparison, one must only compare reactions in which the effective Franck-Condon factor $(\Delta G - \lambda)$ is held constant at $(\Delta G - \lambda) \cong 0$.

Thus, some specific electronic coupling between the donor and acceptor via the intervening matrix seems likely. As an approach to better understanding just how this coupling occurs, Closs[5] (with Morokuma) has undertaken rather detailed molecular orbital calculations involving a bridging cyclohexane unit. When these results are compared with experiment, the agreement is remarkable between the calculated and measured matrix elements ($|V|_{ex}^{calcd} = 10^{-2}cm^{-1}$; $|V|_{ex}^{meas} = |V|_{ex}^{calcd} = 10^{-2}cm^{-1}$!). Such calculations underscore the important role of the intervening matrix in mediating electronic coupling even for an "insulating" bridge like cyclohexane.

2. EFFECTS OF DONOR–ACCEPTOR ANGLE ON ELECTRON TRANSFER

By comparison with the (relative) wealth of theory and experimental data on distance effects, little is known about how the angular orientation of the donor and acceptor affects electron transfer rates. Since wavefunctions have both a radial and an angular component, angle effects must exist. However, their magnitude is uncertain.

In a disordered medium, like the randomly doped glasses used to investigate distance effects, one might anticipate that angle effects would be difficult to observe, since they would be averaged out over all possible angular orientations. Since theory faces no such (practical) obstacles, some important clues as to the relative magnitudes that might be anticipated for such effects are available from theoretical studies.

There have been several theoretical approaches toward treating the effects of donor–acceptor angle. A particularly ambitious ab initio study by Closs and Morokuma considered the effect of *syn* and *anti* substitution on a cyclohexane ring of the STO3G level.[5] Small effects (~5× in rate) were predicted, in good agreement with observation. It should be emphasized that these calculations explicitly consider through bond coupling, so that not only is the overall angle of the donor to the acceptor important, but also the angles subtended by each element (e.g., carbon atom) in the spaces are important. Thus, one might expect that a *trans anti* periplanar spacer would serve as a more efficient "wire" than would a spacer of equal length but with a *cis* geometry. This prediction seems to be supported by the available data of Closs et al.

A different approach, which allows calculations for more complex molecules, has been pursued by Cave et al.[6] They considered a system

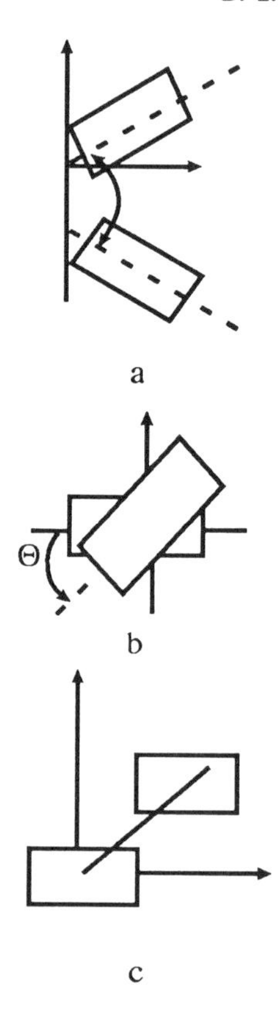

Figure 4. Geometrics Considered by Cave et al.[6] in Their Calculations of Heme to Heme Electron Transfer Rate.

of two porphyrins with *no* covalent bridge, in several different geometries (Figure 4).

Relative matrix elements were calculated using a much simpler extended Hückel method, which, by emphasizing the noted structure of the π system wavefunction, is better suited for gaining qualitative insight for this rather complex molecular system. These calculations make some initially surprising predictions. For example, in case b, the minimum matrix element (and thus minimum rate) does *not* occur when the porphyrin rings are orthogonal, at 90°. Rather, the rates at 90° and 0° are identical, while the minimum rate is predicted at 45° (Figure 5). This surprising result can be understood by a simple "group theory" analysis.

Our lab and others have tried to address some of these questions by the synthesis and study of biomolecular systems with well-defined

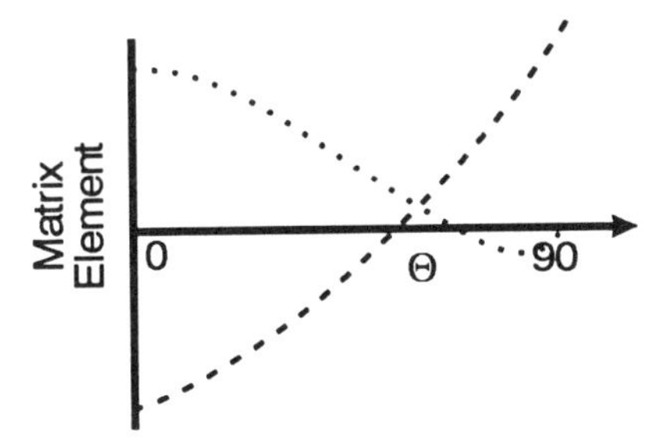

Figure 5. Schematic of the Dependence of Rate on Angle for Two Porphyrins as in Figure 4b .

angular distributions resulting from the steric constraints of the bridging group. Examples of such molecules are shown in Figure 6. By varying the substituent or pattern on the (phenylene) spacer, as indicated in Figure 6, it is possible to systematically vary the porphyrin → porphyrin dihedral angle: Θ. The results of our own studies are shown in Figure 7.

The apparent agreement between the theory of Cave, Siders, and Marcus[6] and the experiments is striking. When the relative simplicity of the theory is considered and compared with the many complications in the experiments, such agreement goes beyond striking toward unbelievable. Indeed, Closs has pointed out[8] a possible alternate explanation that focuses on the unique characteristics of the biphenylene bridge. The essence of the argument is illustrated in Figure 7. As the donor–acceptor

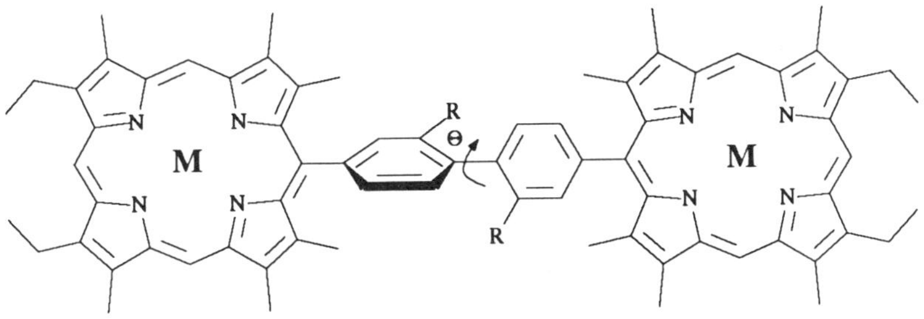

Figure 6. Bisporphyrin Molecules as Synthesized in Cave et al.[6] for Studies of the Angle Dependence of Electron Transfer.

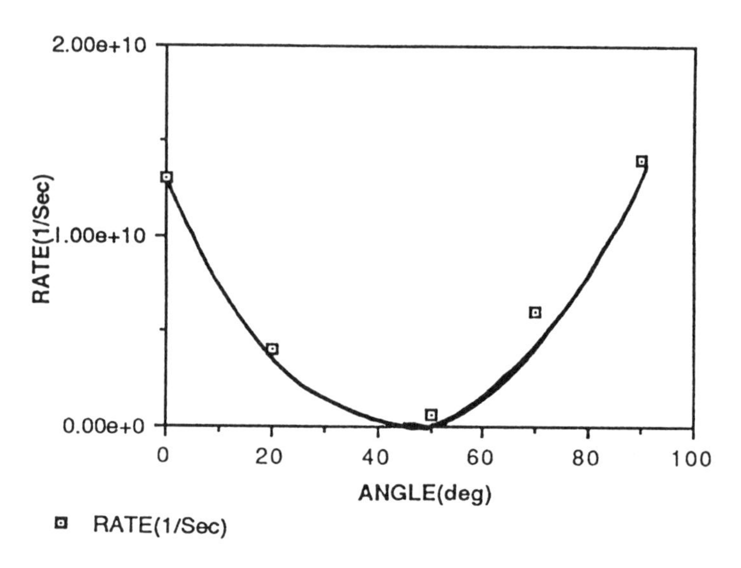

Figure 7. Observed[7] Dependence of Rate on Angle in a Bisporphyrin System. The Solid Line is the Predicted Dependence[6] Scaled to Set $k_{0°} = k_{obs}$.

angle varies, the efficiency of through bond coupling also varies, as shown, with maxima at 0° and 90° and a minimum (node) at 45°.

Thus, two different approaches equally explain the data even though they make different basic assumptions (Marcus assumes *no* through bond coupling, Closs assumes *only* through bond coupling). However, both assumptions rely on basic group theoretic treatments of the nodal structure of the wavefunctions of the donor, acceptor (and spacer). These results minimally predict that there will be no "universal angle dependence": each molecular donor and acceptor has its own nodal structure that will in turn govern the donor–acceptor combinations that are bonding and nonbonding. Thus, the dependence observed for a porphyrin-to-porphyrin transfer will be very different from that observed for a porphyrin-to-quinone transfer, which again is different from a metal ion-to-metal ion transfer. (The near spherical symmetry of simple octahedral metals suggests there should be very little angle dependence for electron transfer between simple metal ions.) A coherent and general theoretical strategy has yet to be developed, and there remain *very* few experimental studies of angle dependence.

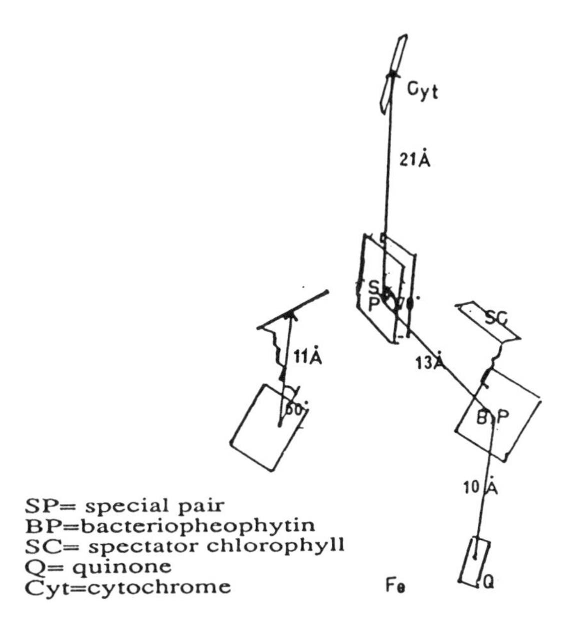

Figure 8. Schematic of the Photosynthetic Reaction Center (after Epp et al.[9]) Indicating the Relative Orientations of the Reactive Centers.

3. APPLICATIONS IN BIOLOGICAL ELECTRON TRANSFER

3.1. The Photosynthetic Reaction Center

The structure of the photosynthetic reaction center as determined by Michl and co-workers[9] is shown in schematic form in Figure 8, which emphasizes the relevant distances and angles of the key pigments.

Although there is some residual controversy, it seems generally agreed that there is not direct electron transfer between the special pair (1) and the spectator chlorophyll (2), i.e., (2) never develops any radical ion character. However, it is equally clear that direct electron transfer between (1) and the phaeophytin (3) occurs at a quite fast rate, $k_{1\rightarrow 3} \approx 10^{11}\ s^{-1}$, suggesting a rather strong coupling ($\geq 50\ cm^{-1}$) over a distance of 10 Å. In terms of the "exponential" distance dependence already discussed, we would require $\exp(-\alpha R)$, $\alpha \leq 0.2$ Å! Clearly something very special occurs to facilitate electron tunneling. The obvious candidate for this special interaction is the "spectator" chlorophyll (2). It has been suggested that this accessory pigment facilitates ($1 \rightarrow 3$)

coupling via the "superexchange" mechanism already discussed. Some evidence in model systems for such superexchange has been provided by Sessler et al.[10] The angles subtended between 1, 2, and 3 are also quite consistent with the superexchange picture.

3.2. Other Biological Systems

3.2.1. *Multichromophore Electron Transport Proteins*

Some of the most interesting electron transport proteins like cytochrome *b* or cytochrome oxidase are involved in proton translocation,[11] in which the physical placement of the chromophores must play a critical role. It is unfortunate that these enzymes have unknown structures. As they are membrane bound, multisubunit systems, they will require clever breakthroughs in crystallization techniques analogous to those used by Michl in the reaction center crystallization before a crystal structure can be obtained.

Thus, our present knowledge of multicenter systems is quite limited. An interesting "model" system is provided by electron exchange within the hemoglobin tetramer. In an elegant and ongoing series of experiments, Hoffman and co-workers examined photoinduced electron transfer between a hybrid[12] tetramer in which 2 Fe hemes have been replaced by photoactive Zn porphyrins to produce the $\alpha Zn_2/\beta Fe_2$ hemoglobin hybrid.

In this hybrid, the closest heme–heme distance is ~20 Å, as demonstrated crystallographically. Rates of ~100 s^{-1} for the reaction $\alpha^3Zn^*\beta Fe(III) \rightarrow \alpha Zn^{\bullet +}Fe(II)$ have been measured; the recombination rate $\alpha Zn^{\bullet +}\beta Fe(II) \rightarrow \alpha Zn\beta Fe(III)$ is somewhat faster at ~400 s^{-1}. From studies of both temperature dependence and free energy variations, the reorganization energy has been estimated as $\lambda \sim 1.2$ V, giving an effective exponential damping constant, $\alpha = 1.0$ Å^{-1} (for $R = 24$ Å; $\nu_0 = 10^{15}$ s^{-1}).

No other multichromophoric system has been studied in equivalent detail. Two such systems of known structure include the yeast lactate dehydrogenase (cytochrome b_2), a heme-flavo enzyme under study both in our laboratory and in France, and the recently characterized ascorbate oxidase, a multiprotein redox enzyme. In our view, such systems represent frontier areas for investigation of how biological systems modulate basic physicochemical parameters.

3.2.2. (Dynamic) Protein to Protein Electron Transfer Complexes

Finally, we briefly consider a third class of biological electron transfer, that occurring between two different proteins. As a first step in such a transfer, a (transient) complex must form, which brings the electron donor and acceptor sites into close proximity so as to maximize the electron tunneling probability. This requirement limits the range of possible productive complexes: a "face to face" approach will work; "back to back" will not (see Figures 9 and 10). This molecular recognition between proteins also ensures specificity of electron transfer, thus avoiding deadly biological "short circuits." The first approaches to understanding recognition focused on rather stereospecific models for complementary charge interactions. An example of such a model, for the well-studied cytochrome *c*:cytochrome *c* peroxidase complex, is shown in Figure 10.

The gross importance of charge complementarity in binding has been confirmed by studies of ionic strength effects on binding,[13] as well as chemical modification studies.[14] More recent studies, however, have suggested that highly stereospecific "lock and key" recognition is *not* involved in recognition between (many) electron transfer proteins.[15] Instead, binding can occur over rather large "sticky" patches that include both charged and hydrophobic residues. Once bound, rather large-scale (two-dimensional) diffusion with rms displacement >6 Å occurs along the proteins' surfaces, permitting an optimal geometry for electron transfer to be attained.

Thus, for protein-to-protein electron transfer, it is rather difficult to precisely define the electronic coupling since this changes in a time-dependent manner concomitant with protein diffusion. In retrospect, it seems plausible that biological systems might adopt a relative non-stereospecificity for molecular recognition in electron transport. As constituted, the system is highly flexible, so a single "mediator" protein like cytochrome *c* can interact with a variety of partners (fulfilling what Darwin called the "principle of maximum utility"). Were highly stereospecific binding used, the range of possible biologically useful reaction partners would be highly circumscribed.

Finally, as we have shown, although rates depend strongly on distance, the dependence on angles is rather weaker, so that, for two concave (~spherical) proteins, a range of roughly equivalent distances (and rates) can be accessed at several different points along the protein-to-protein contact surface.

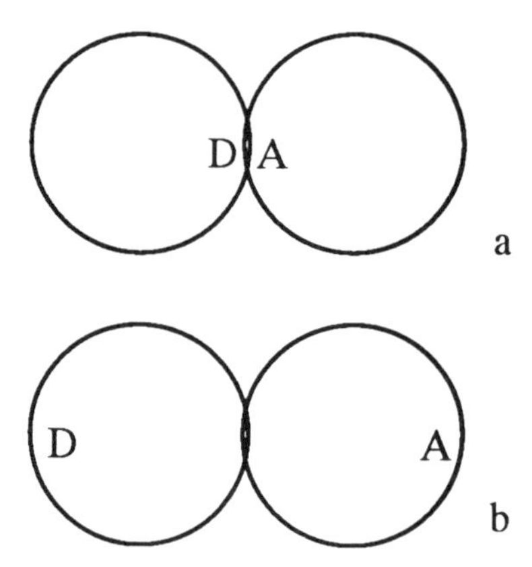

Figure 9. Schematic of Protein-Protein Approach to Form a Precursor Complex for Electron Transfer. "a" is Productive; "b" is Not.

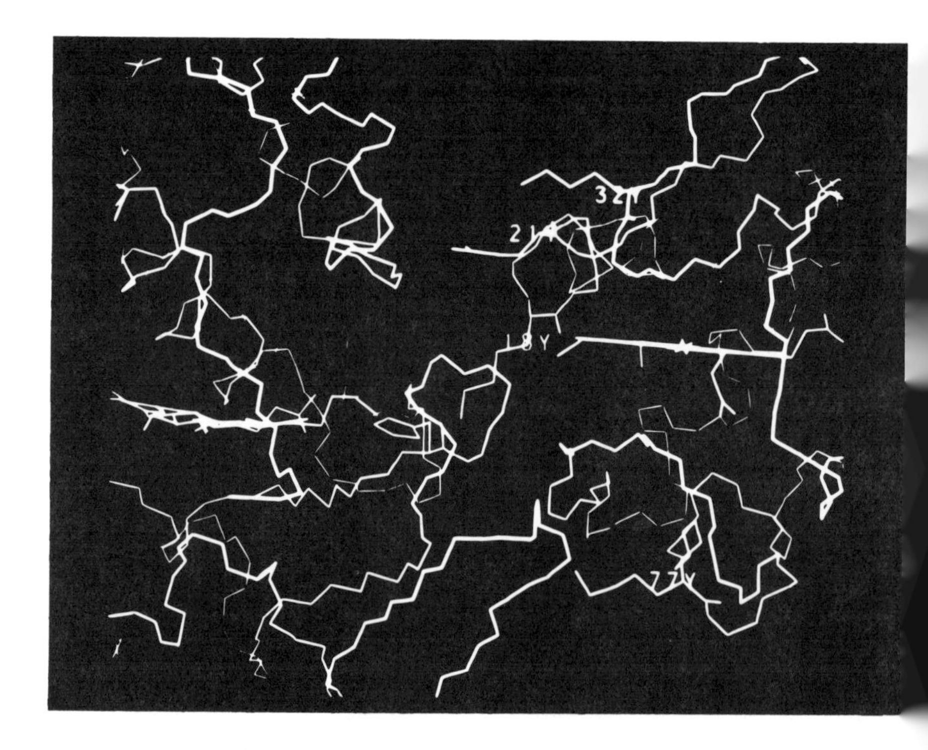

Figure 10. Model Structure Proposed by Poulos for the Cytochrome *c* : Cytochrome *c* Peroxidase Complex.

REFERENCES

1. (a) Miller, J. R. *Science* **1975**, *189*, 221. (b) Miller, J. R. In *Tunneling in Biological Systems*, B. Chance (Ed.) Academic Press, New York, 1978.
2. Calcaterra, T.; Closs, C.; Miller, J. R. *J. Am. Chem. Soc.* **1984**, *106*, 3047.
3. (a) Gamow, G. *Z. Phys.* **1938**, *51*, 204. (b) Keitner, N.; Logan, J.; Jortner, J. *J. Phys. Chem.* **1974**, *78*, 2148. (c) Hopfield, J. J. *Proc. Natl. Acad. Sci. USA* **1974**, *71*, 3640. (d) Beitz, J.; Miller, J. R. *J. Chem. Phys.* **1981**, *74*, 6746.
4. (a) Guarr, T.; McGuire, M.; McLendon, G. *J. Am.Soc. Chem.* **1985**, *107,* 5104. (b) Miller, J. R.; Krongauz, V. in preparation.
5. (a) Closs, G. *Science* **1988**, *240*, 440. (b) Closs, G.; Miller, J. R. *J. Phys. Chem.* **1986**, *90*, 3673.
6. Cave, R.; Marcus, R.; Siders, P. *J. Phys. Chem.* **1986**, *90*, 1436.
7. (a) Heiler, D.; McLendon, G.; Rogalskyj, P. *J. Am. Chem. Soc.* **1987**, *109*, 604. (b) Helms, A.; Heiler, D.; McLendon, G. *J. Am. Chem. Soc.*, in press. (c) Mataga, N. In *Electron Transfer in Organic, Inorganic, and Biological Systems*, Bolton, J.; McLendon, G.; Mataga, N. Eds. ACS Symposium Series, American Chemical Society, Washington, D.C., 1990.
8. Closs, G. personal communication.
9. Epp, J.; Michl, H.; Diesenhoffer, M.; Huber, R. *J. Mol. Biol.* **1984**, *160*, 371.
10. Sessler, J.; Johnson, M.; Lin, T.; Creger, S. *J. Am. Chem. Soc.* **1988**, *110*, 3659.
11. (a) Boxer, S. In *Electron Transfer in Organic, Inorganic, and Biological Systems*, Bolton, J.; McLendon, G.; Mataga, N., Eds. ACS Symposium Series, American Chemical Society, Washington, D.C., 1990. (b) Williams, R. J. P. *Chem. Scripta* **1989**, *28A*, 5.
12. Hoffman, B. M. In *Electron Transfer in Organic, Inorganic, and Biological Systems*, Bolton, J.; McLendon, G.; Mataga, N., Eds. ACS Symposium Series, American Chemical Society, Washington, D.C., 1990.
13. Mauk, M.; Mauk, A. G. *Biochemistry* **1982**, *21*, 1843.
14. Bosshard, R.; Margoliash, E. *Trends Biochem. Sci.* **1986**, *8*, 316.
15. (a) McLendon, G. *Struct. Bonding*, in press. (b) Williams, R. J. P. *Chem. Scripta* **1989**, *29A*, 63. (c) Rogers, K.; Prochapsky, T.; Sligar, S. *Science* **1988**, *240*, 1657.

ELECTRON TRANSFER REACTIONS FOLLOWED BY RAPID BOND CLEAVAGE:

INTRA-ION PAIR ELECTRON TRANSFER OF PHOTOEXCITED CYANINE BORATES AND CHEMICALLY INITIATED ELECTRON-EXCHANGE LUMINESCENCE

Gary B. Schuster

1. Introduction 164
2. Electron Transfer in Photochemical Reactions 165
    2.1. The Driving Force ΔG_{et}. 167
    2.2. Orbital Interaction at the Transition Structure: Adiabatic and Nonadiabatic Electron Transfer. 168
    2.3. The Reaction Coordinate for Electron Transfer Reactions: Internal and External 170
    2.4. A Time-Course Scenario for Electron Transfer Reactions . . . 172
    2.5. Theories That Relate ΔG_{et} to the Rate Constant for Electron Transfer: Marcus Theory 174

Advances in Electron Transfer Chemistry,
Volume 1, pages 163–197.

ISBN: 1-55938-167-1

2.6. The Experiments of Rehm and Weller: Electron Transfer Quenching of Electronically Excited States 178
2.7. Inverted Region Revealed: Charge Shift and Charge Recombination Reactions 182
3. Electron Transfer Reactions That Involve Bond Cleavage 182
3.1. Intra-Ion Pair Electron Transfer in Cyanine Borates 182
3.2. N,N′-Dimethyldimethylindocarbocyanine Hexafluorophosphate ($Cy^+PF_6^-$): Structure and Physical Properties 183
3.3. Alkyltriphenylborates: Chemical and Physical Properties . . . 184
3.4. Cyanine Borates: Absorption and Ion Pairing in Solution . . . 184
3.5. Laser Flash Photolysis 185
3.6. Intra-Ion Pair Electron Transfer: Estimation of the Free Energy (ΔG_{et}) 186
3.7. Fluorescence Spectroscopy Gives k_{et} 186
3.8. A Mechanism for Intra-Ion Pair Electron Transfer of Cyanine Borates 187
3.9. Chemically Initiated Electron-Exchange Luminescence (CIEEL) 190
3.10. CIEEL from Dimethyldioxetanone: Endothermic Electron Transfer 193
3.11. CIEEL from Dimethyldioxetanone: Exothermic Back Electron Transfer 195
4. Conclusions 195
Acknowledgment 196
References 196

1. INTRODUCTION

Photochemistry is a large discipline incorporating concepts from synthetic, mechanistic, and physical chemistry. Nowhere is this overlap stronger than in the investigation of electron transfer processes in photochemical reactions. During the past two decades photoinitiated electron transfer has assumed a role of central significance in applications ranging from the synthesis of new compounds to the evaluation of the detailed dynamics of molecular motion in condensed phases. The importance of photoinitiated electron transfer is recognized in life processes such as photosynthesis and in practical applications of technology such as photoinitiated polymerization reactions. In view of the important role that these reactions play in both theoretical and practical chemistry, it is no surprise that photoinitiated electron transfer has grown to a field of immense size. No single chapter, indeed no single book of manageable weight, can cover even a fraction of the important theories and experimental results that comprise this topic. With this fact in mind, this chapter is restricted at the outset to a review of only a small segment of this field. The other chapters in this book and other available work provide a broader perspective.[1] Herein we will examine primarily electron transfer reactions that result in or are caused by rapid cleavage of a bond in either the donor or the acceptor. Even this field is too large to cover here in great breadth, so this chapter is further restricted and will focus on an analysis of examples of intermolecular electron transfer reactions of organic compounds taken primarily from our work in this area.

From an organizational point of view, we will first examine the theories of electron transfer reactions to define terms of importance. This examination will emphasize those theoretical issues that are of particular relevance for electron transfer reactions that involve bond cleavage. We will then proceed to the investigation of examples of this sort of electron transfer process. The first example comes from the intra-ion-pair electron transfer reaction of cyanine borates. Here electron transfer from the anionic borate to the electronically excited cationic dye results in cleavage of a carbon–boron bond and the formation of a free radical. This process is of importance in the photoinitiation of polymerization with visible light. The second is taken from the field of chemiluminescence—photochemistry in reverse—where a thermally initiated electron transfer reaction to an organic peroxide results in the cleavage of an oxygen–oxygen bond and, eventually, in the generation of light.

2. ELECTRON TRANSFER IN PHOTOCHEMICAL REACTIONS

Electronically excited states are simultaneously strong one-electron acceptors and strong one-electron donors. In simple terms, this bifunctional character is a consequence of the electron promotion that forms the excited state and creates two singly occupied molecular orbitals. One of these, coming from the highest occupied molecular orbital of the ground state compound, is at relatively low energy and is capable of receiving an electron from a donor having an occupied orbital at higher potential. If the excited state is a neutral compound and the electron donor too is neutral, the most commonly encountered case in organic chemistry, then the electron transfer reaction creates a pair of oppositely charged radical ions; this reaction is illustrated in Eq. 1.

The second singly occupied molecular orbital of the excited state is derived from the lowest unnoccupied orbital of the ground state compound. The electron in this orbital is at relatively high potential. When the excited state encounters an acceptor, a reagent having an unfilled orbital at relatively low energy, transfer of an electron converts the excited state into a radical cation and the acceptor to a radical anion—when both were initially neutral (Eq. 2). The reactions shown in Eqs. 1 and 2 involve two electrically neutral components undergoing an electron transfer to form a radical ion pair. This is termed a charge separation reaction. Charge annihilation reactions are just the opposite: a radical ion pair undergoes electron transfer to give two neutral products, one of which may be electronically excited. A charge shift reaction occurs when an ion or an ion radical transfers an electron to a neutral compound.

$$A + D^* \rightarrow (A^{\bullet -}\ D^{\bullet +}) \text{ ion pair} \quad (1)$$

$$A^* + D \rightarrow (A^{\bullet -}\ D^{\bullet +}) \text{ ion pair} \quad (2)$$

In principle the driving force for an electron transfer reaction (ΔG_{et}) may be readily calculated from experimentally measurable parameters such as the oxidation potential (E_{ox}) of the donor, the reduction potential of the acceptor (E_{red}), the energy of the excited state (ΔE^*), and Coulombic energy terms associated with separation or recombination of charge. One of the major goals in the examination of electron transfer reactions is to relate ΔG_{et} to the rate constant for the process (k_{et}). Theories of various sophistication have been developed and advanced to relate these parameters. In this section we will examine the issues associated with estimation of ΔG_{et} with particular emphasis on electron transfer reactions that result in bond cleavage. This is a complex case that is sometimes unappreciated. Then we will review briefly the relationship between ΔG_{et} and k_{et} embodied in classical Marcus theory, the approach of Rehm and Weller, and the recent trifurcation of Mataga and Kakitani for charge separation, annihilation, and shift reactions. Finally, we will briefly review two recent successful demonstrations of the Marcus theory of electron transfer.

2.1. The Driving Force ΔG_{et}.

It is informative to consider first the free energy change in an electron transfer reaction between ground state reagents in the gas phase. Here definitions are clear:

$$\Delta G_{et} = I_D - E_A - e^2/r_{AD} \tag{3}$$

where I_D is the ionization potential of the donor and E_A is the electron affinity of the acceptor. These are parameters with well-defined thermodynamic meaning unencumbered by the complexity of medium effects. The third term on the right side of Eq. 3 accounts for the Coulombic energy when both the acceptor and donor are neutral and form an ion pair at distance r_{AD}—a charge separation reaction. The Coulombic energy term is readily calculated in the gas phase.

Electronic excitation of the donor lowers its ionization potential by the excitation energy whereas excitation of the acceptor raises its electron affinity. The effect on ΔG_{et} is the same regardless of which partner is excited—it is decreased by ΔE^*:

$$\Delta G_{et} = I_D - E_A - \Delta E^* - e^2/r_{AD} \tag{4}$$

When the electron transfer reaction takes place in solution, the effect of transferring the donor, acceptor (excited or ground state), the ions $D^{\bullet+}$ and $A^{\bullet-}$ and the ion pair $\{D^{\bullet+}A^{\bullet-}\}$ from vacuum to a medium with dielectric constant ε must be considered. Now E_{ox} and E_{red}, measured electrochemically, replace I_D and E_A according to

$$(I_D - E_A) = [(E_{ox} - E_{red}) + (E_+ + E_-)] \tag{5}$$

where E_+ and E_- are the differences in solvation energy between the neutral molecules and the ions.[2]

There are two complications of significance that must be considered in the application of Eq. 5. The first is that thermodynamically meaningful oxidation and reduction potentials often cannot be obtained. In particular for the case of prime interest in this chapter where rapid bond cleavage follows oxidation or reduction; electrochemical techniques such as cyclic voltammetry give only peak potentials that contain both thermodynamic and kinetic components. This issue will be addressed more completely later. The second complication comes from the estimation of $(E_+ + E_-)$. This is commonly performed by application of the Born equation:

$$(E_+ + E_-) = -e^2/2\{1/r_D + 1/r_A\}\{1 - 1/\varepsilon\} \quad (6)$$

where r_D and r_A are the radii of the presumably spherical ions. This application of the Born equation is often criticized[3] on the basis of its derivation from classical electrostatics and for the often crude assumption of spherical ions.

Finally as in the case of the gas phase electron transfer reaction, the electrostatic term in solution must be considered. The issue here is particularly complex when the electron transfer occurs at contact distance between D and A. In this circumstance it is certainly improper to use the bulk solvent dielectric constant as a screening factor.[4] Nevertheless, the usual solvent correction to Eq. 4 follows Kirkwood's approach[5] in calculating the energy for transfer ($\Delta G_{A^-D^+}$) of an ion pair with dipole moment μ from the gas phase to solution:

$$\Delta G_{A^-D^+} = -\mu^2/r^3\{(\varepsilon - 1)/(2\varepsilon + 1)\} \quad (7)$$

where r is the radius of the ion pair.

These considerations lead to a rather formidable, but still highly compromised, equation for estimation of the free energy change for a photoinitiated electron transfer reaction in solution like those shown in Eq. 1 or 2:

$$\Delta G_{et} = (E_{ox} - E_{red}) - \Delta E^* + (E_+ + E_-) - e^2/r_{AD} - \Delta G_{A^-D^+} \quad (8)$$

If the electron transfer reaction is carried out in a very polar solvent such as acetonitrile, the Coulombic energy term is small and is commonly neglected. However, some of the electron transfer reactions we consider in this chapter are carried out in nonpolar solvents such as benzene. In this circumstance the Coulombic term is large and must be considered, however approximately. Finally in the derivation of Eq. 8, we have ignored any stabilization (or destabilization) of the reagents or products due to specific orbital (ground state complex or exciplex) interactions. In the specific cases we will examine below, there is no direct experimental evidence that suggests the need to consider these factors.

2.2. Orbital Interaction at the Transition Structure: Adiabatic and Nonadiabatic Electron Transfer

In most electron transfer reactions, and in all of the cases examined

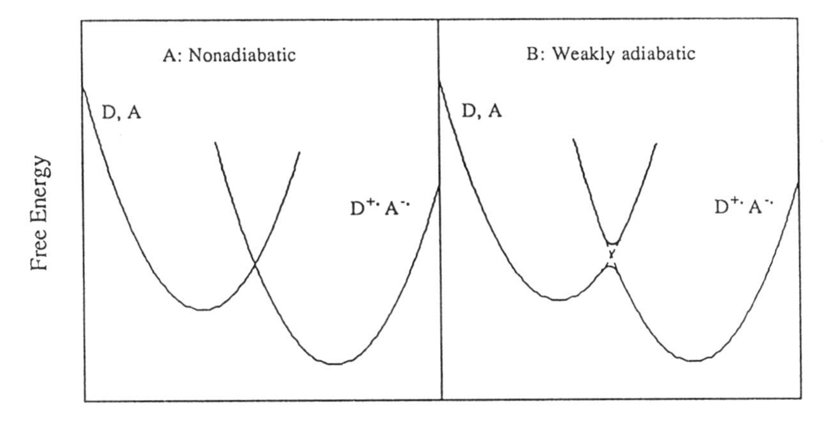

Figure 1. (A) Schematic Representation of a Nonadiabatic Intersection Between Two Potential Energy Surfaces. The Left-Hand Parabola Represents the Starting State for the Reaction, the Right-hand Parabola the Product State. (B) Schematic Representation of a Weakly Adiabatic Electron Transfer Reaction. The Intersection is Weak if the Splitting in the Intersection Region is Appropriately Small with Respect to Normal Thermal Energies.

specifically in this chapter, the donor and the acceptor are nearest neighbors in solution when the electron transfer occurs. There is another category, long-distance electron transfer,[6] that generally involves a donor and acceptor in the same molecule separated by a rigid spacer group so that their orbitals cannot strongly overlap. The strength of the interaction between the orbitals of the donor with those of the acceptor during the electron transfer divides these reactions into two limiting categories: adiabatic and nonadiabatic.

In the nonadiabatic limit, there is no interaction at all between the orbitals of the donor and the orbitals of the acceptor (Figure 1A). If one visualizes the nonadiabatic case as two-dimensional potential energy surfaces in reaction coordinate space, the parabola representing the starting state (D,A say) will intersect the parabola representing the product state ($D^{\bullet+}$,$A^{\bullet-}$). The intersection represents the point where D, A and $D^{\bullet+}$, $A^{\bullet-}$ have the same geometry and the same energy but have different electronic configurations. In the nonadiabatic limit there is no mixing at the intersection point and electron transfer from D to A cannot occur. In reality there probably will be no actual examples of the nonadiabatic limit when the donor and acceptor are polyatomic molecules. In these cases there are too many degrees of freedom available to enforce the strict maintenance of symmetry that the nonadiabatic limit requires. Some weak mixing even when D, A and $D^{\cdot+}$, $A^{\cdot-}$ have different spins, i.e., singlet and triplet states, occurs through the operation of spin orbit coupling.[7]

The more mixing at the intersection point, the more the reaction description moves toward the adiabatic limit. Now there are two additional categories that need to be considered: weakly adiabatic and strongly adiabatic electron transfer reactions. Conceptually, in the weakly adiabatic case, (Figure 1B) there is a finite electronic coupling between the D, A and $D^{\cdot+}$, $A^{\cdot-}$ states at the intersection, but the coupling is "small." In this connection the size of the coupling is measured in relation to the energy difference between the starting state and the transition structure in the nonadiabatic limit. The prime significance of weakly adiabatic case is that electron transfer occurs and its activation energy (rate) can be accurately estimated, in principle, by calculations for the nonadiabatic case. The theories that relate ΔG_{et} to the rate of electron transfer that we will consider below are based on the premise that they are representatives of the weakly adiabatic case.

In reality many electron transfer reactions, particularly those involving electronically excited states, may not conform to the requirement of a small electronic coupling at the transition structure. In these cases there is a strong interaction between the orbitals of the donor and those of the acceptor at the transition structure. The magnitude of this interaction depends on the details of the structure of the donor and acceptor. In this, the strongly adiabatic case, the activation barrier for electron transfer is controlled primarily by the strength of the electronic coupling between D, A and $D^{\cdot+}$, $A^{\cdot-}$ at the transition structure. The classical theories that relate ΔG_{et} to the rate of electron transfer cannot be easily applied in the strongly adiabatic limit.

2.3. The Reaction Coordinate for Electron Transfer Reactions: Internal and External

It is clear from consideration of Figure 1 that internal vibrational coordinates of D, A and $D^{\cdot+}$, $A^{\cdot-}$ control, in part, the position of the intersection point where the electron transfer reaction occurs. Consider Figure 2, which has been drawn so that the parabolas representing the starting and the final states have the same energy ($\Delta G_{et} = 0$). Further, for simplicity, the parabola representing D, A is held constant, but the width

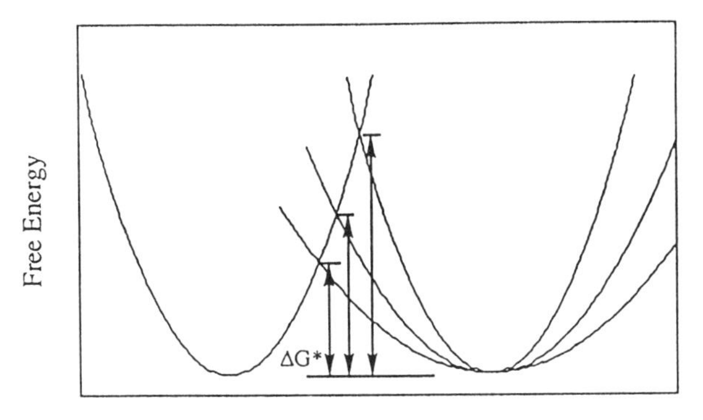

Figure 2. Schematic Representation of the Effect of Changes in the Internal Reorganization Energy on the Activation Barrier for an Electron Transfer Reaction.

of the $D^{\bullet+}$, $A^{\bullet-}$ parabola is widened. In this regard, widening the parabola can be thought of as weakening the force constants for vibrations in the $D^{\bullet+}A^{\bullet-}$ state. As the parabola widens, the point of intersection between the two curves moves to lower energy. This corresponds to a lower activation barrier and a faster electron transfer reaction. In reality it is impossible to change the force constants of the reactants or products significantly without simultaneously affecting ΔG_{et}. Nevertheless the implication is clear, an important component to the description of the reaction coordinate is the positions of the atoms of the donor and the acceptor in space. This contribution to the activation barrier is designated the internal reorganization energy and is given the symbol λ_i.

In theory the magnitude of λ_i may be calculated from first principles. Classically it is given in Eq. 10 where the $k_i^r - k_i^p$ represent the normal mode vibrational force constants in the reactant D, A state and the product $D^{\bullet+}$, $A^{\bullet-}$ states, respectively, and the difference $(q_i^r - q_i^p)$ represents the changes in bond lengths and angles on going from reactants to products. The summation is over all degrees of freedom.

$$\lambda_i = \sum_i \frac{k_i^r \times k_i^p}{k_i^r + k_i^p} [q_i^r - q_i^p] \qquad (10)$$

A second component to the reaction coordinate for electron transfer reactions is the solvent reorganization energy. This parameter is designated λ_s and in some cases it is the dominant factor controlling the rate of electron transfer. It is obvious that the solvent will play an important role in electron transfer reactions of the kind depicted in Eqs. 1 and 2. The starting state for these reactions is nonionic and the particular organization of solvent dipoles around the reactants does not affect the energy of this state very much. On the other hand, the final state, $D^{\bullet+}$, $A^{\bullet-}$ is ionic and its energy will be very sensitive to the orientation of the solvent dipoles. It is also important to consider the polarizability of the solvents in evaluating its response to the charge separation. Both the polarizability and the polarity of the solvent are included in λ_s in Eq. 11, where n is the refractive index of the solvent. Equation 11 is derived from the assumption that the donor and acceptor are spheres with radii r_D and r_A and that the ion pair has radius r_{AD}.

$$\lambda_s = e^2[1/2r_D + 1/2r_A - 1/r_{AD}][1/\varepsilon - 1/n^2] \qquad (11)$$

The total reorganization energy (λ) for the electron transfer reaction is simply the sum of the internal and solvent components (Eq. 12). In classical theories of electron transfer reactions, for the weakly non-adiabatic case, the rate of the reaction can be predicted simply from knowledge of ΔG_{et} and λ.

$$\lambda = \lambda_i + \lambda_s \qquad (12)$$

2.4. A Time-Course Scenario for Electron Transfer Reactions

The transfer of an electron can be viewed as instantaneous. It is clear, however, that electron transfer reactions are not instantaneous; some that we will consider below are rather fast having rate constants of $\sim 10^{12}\ s^{-1}$; others are quite slow, sometimes taking hours to complete. The reconciliation between instantaneous transfer of an electron and variable rate of electron transfer reactions is easily accomplished by realizing that the slow step in the process is getting the system ready to transfer an electron.

It is illustrative to consider the time course of the reactions described in Eqs. 1 or 2. Light is absorbed by either D or A to create an electronically excited state. We presume that there is no ground state interaction between D and A so that at the instant of excitation D and A are completely surrounded by solvent and are isotropically distributed in solution. In our scenario we require that random diffusion eventually

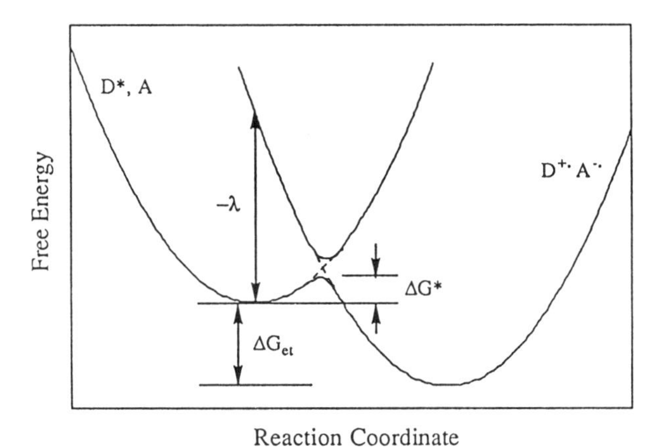

Figure 3. Electron Transfer from an Electronically Excited State D* to a Ground State Acceptor (A). Net Observable Reaction Occurs at the Intersection Region Where the Initial and Final States Have the Same Geometry.

results in an encounter between the excited state and its reaction partner. The probability of this occurring will increase with the lifetime of the excited state, the concentration of the partner, and the diffusion rate (controlled by viscosity). When the excited state and its partner become nearest neighbors both will be almost exclusively in their lowest energy vibrational states and the solvent orientation around the pair will be primarily in its lowest energy orientation. If an electron transfer reaction were to occur under these conditions, it could be represented by the vertical arrow on Figure 3. This is an endothermic process raising the energy of the system by the reorganization energy λ. Recall that transfer of the electron is instantaneous and that the reaction coordinate is comprised of nuclear motion terms (internal and solvent). Nuclear motions occur much more slowly than does the electron transfer, hence the transition is vertical in this reaction coordinate system—it occurs with no change in nuclear coordinates.

As pictured in Figure 3, the vertical electron transfer is endothermic, the $A^{\bullet -}$, $D^{\bullet +}$ state is higher energy than the (A,D)* state. We can consider two limiting options for the high energy charge transferred state. It could relax as the charge transferred state by changes in internal or solvent nuclear coordinates, or it could return to the starting state. In our scenario, the return to the starting state requires only the exothermic transfer of an electron and therefore is instantaneous. Relaxation in the charge transferred state requires time to change nuclear coordinates. When comparing two processes originating from the same state one of which takes time the other being instantaneous, the only one that actually will occur is the instantaneous one. Thus no net electron transfer occurs from (A,D)* when they are in their initial encounter internal and solvent geometry.

Precisely the same considerations apply to all geometries represented in Figure 3 except the one described by the intersection between the surfaces representing (A,D)* and $D^{\bullet+}$, $A^{\bullet-}$. Here the energies and geometries of the two states are the same so electron transfer in either direction is instantaneous. A reaction coordinate change to the right after the intersection stabilizes the electron transferred product state; movement to the left stabilizes the starting state.

With these considerations in mind, the scenario for a photoinitiated electron transfer reaction has three acts. First the excited state and its partner diffuse together to become nearest neighbors. Next internal and solvent motions combine to raise the energy of the starting state to the intersection point where electron transfer occurs. Finally, relaxation of internal and solvent coordinates (for exothermic electron transfer) stabilizes the product state. The rate-determining step in this scenario is the fluctuation in internal nuclear and solvent position that brings the system to the intersection point.

2.5. Theories That Relate ΔG_{et} to the Rate Constant for Electron Transfer: Marcus Theory

A major goal in chemistry has been the attempt to relate the free energy change for a reaction to its rate. An attempt familiar to organic chemists is the linear free energy relation introduced by Hammett.[8] This approach quantifies the common experience that the rate of a reaction increases as it becomes more exothermic. In the linear formalism, the reaction rate is predicted to always increase when the driving force does.

In 1957 Rudolph A. Marcus published a theory, which now bears his name, that showed for the case of electron transfer reactions that the linear free energy relation is an inappropriate description.[9] Marcus theory predicts a quadratic relationship between the driving force of a reaction and its rate. If one considers first reactions that are endothermic, their rates will increase as the driving force does until the rate reaches a maximum value. Further increase in the driving force of the reaction is predicted to result in a decrease in the reaction rate. In its original classical form Marcus theory (Eq. 13) predicts a parabolic relationship between $\Delta G^{\ddagger}$, the activation energy for electron transfer, and ΔG_{et}:

$$\Delta G^{\ddagger} = \Delta G_{es} + \lambda/4[1 + \Delta G_{et}/\lambda]^2 \quad (13)$$

where ΔG_{es} is an electrostatic term reflecting loss or gain in electrostatic free energy as the reagents form an encounter complex. For electron transfer reactions where at least one component is electrically neutral, as is the case for the reactions described in Eqs. 1 and 2, $\Delta G_{es} = 0$.

In the derivation of Eq. 13 all vibrations are treated as classical oscillators. In the semiclassical Marcus approach, the internal vibrations of the donor and acceptor are treated quantum mechanically but the solvent normal modes are viewed as classical oscillators. Fully quantized Marcus theory treats all internal and solvent vibrations quantum mechanically. From the point of view of this chapter, it is necessary to consider only the implications of classical Marcus theory in the analysis of the electron transfer reactions of interest.

Inspection of Eq. 13 reveals that the maximum rate constant for electron transfer occurs when $\Delta G_{et} = -\lambda$. If the driving force for the reaction is greater than this value, the reaction rate is slower than the maximum value. This is referred to as the "inverted" region because it violates the usual experience that as a reaction becomes more exothermic its rate increases. If ΔG_{et} is less than $-\lambda$, then increasing the reaction exothermicity results in an increase in the reaction rate; this is called the "normal" region since it is consistent with our usual experience. It is possible to reach a qualitative pictorial understanding of the Marcus normal and inverted regions by examining the intersection point of two identical, but displaced, parabolas.

The parabolas on the left in Figure 4 represent the starting electronic state of the system, D, A; the curves on the right are the final state $D^{\bullet +}$, $A^{\bullet -}$. The two curves in Figure 4a are drawn so that $\Delta G_{et} = 0$. For this case $\Delta G^{\ddagger} = \lambda/4$. It is clear from inspection of Figure 4A that raising the right-hand curve relative to the one on the left (i.e., increasing ΔG_{et}) results in an increase in $\Delta G^{\ddagger}$. This is the normal region. The curves in Figure 4B are drawn by lowering the parabola representing the products—making the reaction more exothermic—so that in this case $\Delta G_{et} = -\lambda$. The intersection between the two curves now occurs at the bottom of the curve for the starting state: there is no activation barrier for the reaction; it occurs at its maximum rate. An additional increase in the driving force for the reaction is shown in Figure 4C. Now the reaction is

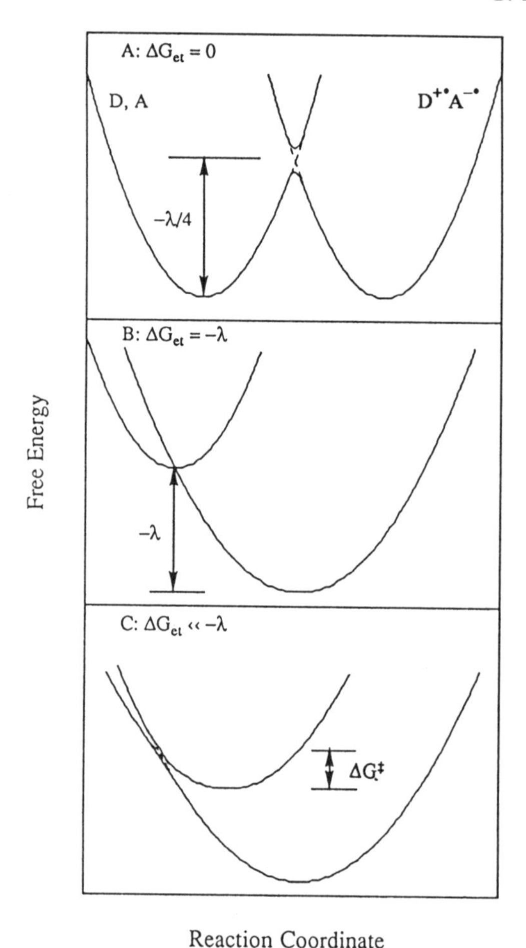

Figure 4. (A) Activated Electron Transfer in the Normal Region When $\Delta G_{et} = 0$. (B) The Maximum Rate for Electron Transfer, $\Delta G_{et} = -\lambda$. Electron Transfer in the Inverted Region; Note that There is No Intersection Between the Curves Representing the Starting and the Final States.

very exothermic, the product state is much lower in energy than the starting state, and the point of intersection is moved to the left side of the parabola describing the starting state. This is the inverted region: additional increase in the driving force results in an additional increase in $\Delta G^{\ddagger}$ and a slower rate of reaction.

Though it is possible to achieve a pictorial understanding of the origin of the Marcus inverted region from analysis of intersecting parabolas, a more fundamental understanding requires consideration of quantum mechanical energy transfer events. An in-depth analysis of this topic is beyond the purview of this chapter. However, it is informative to consider the origin of the Marcus inverted region in connection with the well-known and easily observed inverse energy-gap rule for the radiationless deactivation of electronically excited states.[10]

Electronically excited states can relax with the emission of light—a radiative transition—or by conversion of the energy of electronic excitation to heat—a radiationless transition. In the latter process the energy of the excited state (potential energy) is transferred to vibrational modes of the ground state (kinetic energy). One can consider for the purpose of illustration that the electronically excited state is in its lowest vibrational state (classically) moving over only a small displacement in nuclear coordinates. The radiationless transition occurs instantaneously: in one instant the molecule is electronically excited with low internal kinetic energy; in the next it is in its electronic ground state with high internal kinetic energy. In the newly formed ground state the atoms must suddenly begin to move with relatively large displacement in their coordinates. Thus the radiationless transition requires the instantaneous conversion of potential energy to kinetic energy. It is not difficult to see that in such a process the greater the amount of potential energy converted to kinetic energy, the less likely the conversion will be: a system at rest must instantaneously begin to suddenly move more and more rapidly the greater the amount of potential energy. This concept is expressed quantitatively as the Franck–Condon factor, which is a measure of the overlap of the vibrational wavefunctions of the ground and excited state. The greater the energy released in the radiationless conversion, the smaller the Franck–Condon factor and the slower the process will be. Hence the inverse energy gap law.

The analogy between the inverse energy gap law and the inverted electron transfer region is clear. In the inverted region we are again faced with the conversion of a large amount of potential energy to kinetic energy and thus we anticipate a small Franck–Condon factor that decreases further as the driving force of the reaction increases. Historically, the decrease in the rate of radiationless deactivation of electronically excited states was observed experimentally long ago. However, the certain experimental manifestation of the inverted region for electron transfer reactions was just recently obtained. Indeed, as of this writing there is no existing experimental evidence for inverted region behavior for reactions of the type depicted in Eqs. 1 and 2. We will examine this situation in some detail below

2.6. The Experiments of Rehm and Weller: Electron Transfer Quenching of Electronically Excited States

In 1969 Rehm and Weller reported the results of experiments in which the lifetime of electronically excited singlet states in acetonitrile solution was monitored in the presence of increasing concentrations of quenchers.[11] These results were interpreted in terms of the reaction sequence shown in Scheme 1 where the collision between the excited acceptor molecule and the donor results in the formation of a solvent-separated radical ion pair. This is an electron transfer reaction of the type identified in Eq. 2.

On the basis of Marcus theory, one expects that as ΔG_{et} for this reaction becomes increasingly negative that k_{et} should first increase, when the values are in the normal range, then begin to decrease as the driving force for the reaction enters the inverted region. In the event, the normal region was clearly in evidence but there was no hint of a decrease in the rate constant for electron transfer even when $\Delta G_{et} = -60$ kcal/mol. Rehm and Weller observed that $k_{et} \approx 10^9\ M^{-1}\ s^{-1}$ when $\Delta G_{et} = 0$ and increases to a maximum value of $2 \times 10^{10}\ M^{-1}\ s^{-1}$ (a limit imposed by the rate of diffusion in acetonitrile solution) when $\Delta G_{et} \approx -10$ kcal/mol. The measured rate constant for electron transfer remained at the diffusion controlled limit for all values of ΔG_{et} less than -10 kcal/mol. Rehm and Weller derived an empirical fit to their experimental results that has become known as the Weller equation (Eq. 14) and Eq. 15:

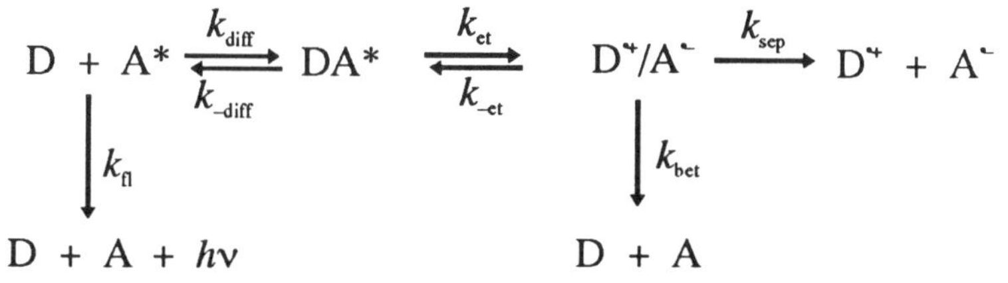

$$\mathrm{D + A^*} \underset{k_{-diff}}{\overset{k_{diff}}{\rightleftharpoons}} \mathrm{DA^*} \underset{k_{-et}}{\overset{k_{et}}{\rightleftharpoons}} \mathrm{D^{\cdot+}/A^{\cdot-}} \xrightarrow{k_{sep}} \mathrm{D^{\cdot+} + A^{\cdot-}}$$

$$\mathrm{D + A^*} \xrightarrow{k_{fl}} \mathrm{D + A} + h\nu \qquad \mathrm{D^{\cdot+}/A^{\cdot-}} \xrightarrow{k_{bet}} \mathrm{D + A}$$

Scheme 1.

$$k_{et} = \frac{2 \times 10^{10}}{1 + 0.25\left[\exp\left(\frac{\Delta G_{et}^*}{RT}\right) + \exp\left(\frac{\Delta G_{et}}{RT}\right)\right]} \tag{14}$$

$$\Delta G_{et}^* = \left[\left(\frac{\Delta G_{et}}{2}\right)^2 + [\Delta G_{et}^*(0)]^2\right]^{1/2} + \frac{\Delta G_{et}}{2} \tag{15}$$

where $\Delta G_{et}(0)$, the activation energy when $\Delta G_{et} = 0$, is assigned the constant value of 2.4 kcal/mol in acetonitrile solution.

The approach of Rehm and Weller has proved very useful in a practical sense since it permits the prediction of the rate constant of a photoinitiated electron transfer reactions in solution from easily obtained thermodynamic parameters. Alternatively, if the rate constant for a photoinitiated electron transfer in the normal region is known, the Weller equation permits the calculation of ΔG_{et}. Below we will employ Eqs. 14 and 15 in the latter sense to estimate the oxidation potentials of compounds that undergo rapid bond cleavage after electron transfer so that their oxidation potentials cannot be measured by electrochemical methods.

The experiments of Rehm and Weller touched off a debate that continues to the present about why the expected inverted Marcus region behavior was not observed. Several explanations for the missing inverted region have been offered.

Weller proposed that in some cases the electron transfer reaction depicted in Scheme 1 does not give the electronic ground state of the radical ions but instead forms them in an electronically excited state. In this scenario ΔG_{et} will not be as negative as anticipated and hence might not fall in the inverted region. There are some recent experiments in support of this general proposal for special cases.[12] Also in the examination of chemiluminescent reactions we describe below there is additional evidence to support this possibility. However, the formation of excited state products cannot be the entire explanation for the absence of the inverted region in the Rehm–Weller experiments since, as Weller points out, the lowest excited state ion pair has an energy of 25 kcal/mol, so inverted region behavior should have been detected when $-25 < \Delta G_{et} < -10$ kcal/mol.

Bixon offered a second explanation for the lack of inverted region behavior in the Rehm–Weller experiments.[13] They suggest that vibrationally excited states of the products are formed by "allowed" Franck–Condon interactions. This would lead to increased splitting at the transition structure and, perhaps, to violation of the weakly adiabatic requirement upon which Marcus theory is based.

Another possibility is that electron transfer occurs at distances greater than contact in the Rehm–Weller experiments. In this circumstance λ will increase and inverted region behavior will be pushed off to more negative ΔG_{et}. Indeed, one could imagine that electron transfer will always be "fast" regardless of the driving force in this scenario. The reaction simply occurs at greater distance the more negative is ΔG_{et}. However, electron transfer occurring at greater than contact distances should give rise to values of k_{et} greater than the diffusion limit. No such values were obtained by Rehm and Weller.

Exciplex formation between the electronically excited acceptor and the donor might also obscures inverted region behavior. In this context we can view exciplex formation within the Marcus framework. Formation of an exciplex requires a strong interaction between the donor and the acceptor. This interaction could make the reaction coordinate appear to be adiabatic. In this circumstance, no inverted behavior is expected.

Another possibility is that quantum tunneling obscures the inverted region in the Rehm–Weller experiments. The tunneling probability is expected to be greater in the inverted than in the normal region because the barrier is narrower in the latter. Nevertheless, tunneling of heavy nuclei seems an unlikely possibility.

Finally, in a recent series of papers Kakitani and Mataga have suggested that the absence of inverted region behavior in the Rehm–Weller experiments is a consequence of dielectric saturation for the special case of charge separation reactions.[14] This proposal nicely accommodates the observation, discussed briefly below, of inverted region behavior by Miller et al.[15] for charge shift reactions and by Gould et al.[16] in the case of charge recombination reactions.

The dielectric saturation effect postulated by Kakitani and Mataga requires different first solvent shell vibrational frequencies that depend on the charge state of the reactants. For neutral reactants the solvent motion is facile and the parabola representing the reactants is relatively shallow. This is qualitatively understood as meaning that the energy of the neutral reactants is not strongly affected by the orientation of the solvent dipoles. In contrast, the solvent is tightly ordered around the ionic product state in a charge separation reaction. Consequently the parabola

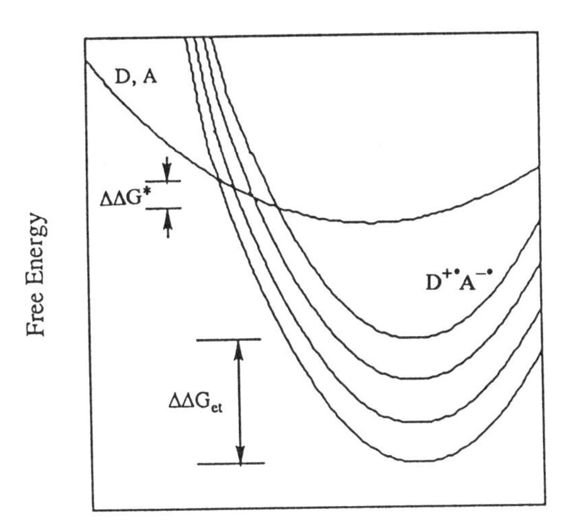

Figure 5. Effect of Solvent Dielectric Saturation on the Activation Barrier for a Charge Separation Electron Transfer in the Inverted Region. Note that a Large Change in Driving Force Causes a Small Change in Activation Barrier.

representing this state is sharp—small changes in solvent dipole orientation greatly influence the energy of this state.

Application of this concept to charge separation reactions (Eqs. 1 or 2) where the starting state is uncharged—a shallow parabola—while the product state is an ion pair with tightly ordered solvent molecules—a sharp parabola—leads to the prediction that charge separation electron transfer reactions with large negative values of ΔG_{et} will occur with near the maximum rate constant. This can be represented pictorially in Figure 5 where the shallow parabola is exaggerated for clarity. It can be seen by inspection that a large increase in the driving force results in only a small change in activation barrier, which, in the inverted region, is always close to its minimum value.

Very recently Marcus examined application the Kakitani–Mataga model to charge transfer absorption and fluorescence spectra.[17] The first process corresponds to a charge separation electron transfer and the latter to a charge recombination reaction. By analyzing the widths of these spectra, Marcus concludes that there is no evidence for dielectric saturation in either reaction. Also, Tachiya recently took a view contrary to that of Kakitani and Mataga about the importance of dielectric saturation in charge shift electron transfer reactions.[18]

In summary, no completely satisfying explanation for the absence of inverted region behavior for the Rehm–Weller experiment has yet been forthcoming.

2.7. Inverted Region Revealed: Charge Shift and Charge Recombination Reactions

In 1979 Beitz and Miller reported charge shift electron transfer reactions in rigid, frozen matrices studied by pulsed radiolysis techniques.[19] In this experiment, trapped electrons were captured by electron acceptors having a range of reduction potentials. Highly exothermic reactions showed an exothermic rate restriction of 100,000-fold—definite experimental proof of the existence of the Marcus inverted region.

Miller et al. also observed the Marcus inverted region by means of a pulse radiolysis experiment in fluid solution at room temperature.[15] In these experiments the donor (biphenyl) and the acceptor were in one molecule kept apart by a rigid saturated hydrocarbon spacer. Radiolysis generates a nonequilibrium distribution of radical anions. Intramolecular charge shift electron transfer from the biphenyl radical anion to the acceptor is the spontaneous direction. By varying the reduction potential of the acceptor, Miller et al.[15] were able to show a bell-shaped dependence of k_{et} on ΔG_{et}. Moreover by varying the solvent used in the experiment, they showed that the rate constants for electron transfer depend on λ_s in the manner predicted by Marcus.

Finally, in a recent series of experiments Gould et al. demonstrated inverted region behavior in a charge recombination electron transfer reaction.[16] In this case the competition between escape of a pair of photogenerated radical ions and their annihilation to reform ground state reagents was measured. The driving force for the reaction was varied in a systematic manner by changing the nature of the electron donor and the electron acceptor. These experiments clearly reveal inverted region behavior and support Marcus theory in additional detail.

3. ELECTRON TRANSFER REACTIONS THAT INVOLVE BOND CLEAVAGE

3.1. Intra-Ion Pair Electron Transfer in Cyanine Borates

We investigated the photochemistry of borate salts of dimethylindocarbocyanine [(Cy^+) $RB(Ph)_3^-$] dissolved in nonpolar solvents. Under these conditions irradiation of the cyanine dye with visible light results in efficient sensitization of the borate and the generation of free alkyl radicals (Eq. 16). This reaction is useful for the initiation of photopolymerization of vinyl monomers with visible light. The examination of the mechanism of the cyanine borate photoreaction by time resolved spectroscopy and by conventional chemical analysis reveals an intra-ion pair charge recombination electron transfer reaction rendered irreversible by the rapid cleavage of the alkyl carbon–boron bond of the boranyl radical.

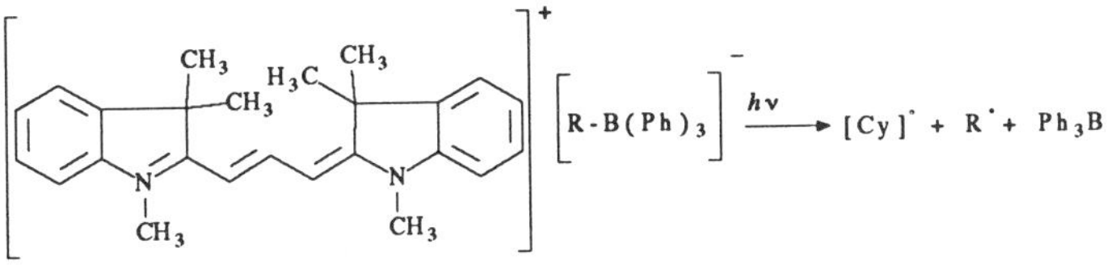

$$[Cy]^+ \, [R\text{-}B(Ph)_3]^- \xrightarrow{h\nu} [Cy]^{\bullet} + R^{\bullet} + Ph_3B \qquad (16)$$

3.2. N,N′-Dimethyldimethylindocarbocyanine Hexafluorophosphate ($Cy^+PF_6^-$): Structure and Physical Properties

Comparison of the electronic absorption spectra of $Cy^+PF_6^-$ dissolved in benzene at room temperature and in a frozen solution at 77 K shows no changes in band shape or position that would indicate participation by more than one conformation of the dye. At room temperature in benzene solution, $Cy^+PF_6^-$ fluoresces with modest efficiency: $\Phi_{fl} = 0.047 \pm 0.01$. The quantum efficiency for fluorescence is independent of excitation wavelength throughout the visible absorption band. The time dependence of the fluorescent emission from benzene solutions of $Cy^+PF_6^-$ at room temperature shows a single first-order decay with a lifetime of 260 ± 10 ps. The absorption and fluorescence data reveal a O–O absorption band for $Cy^+PF_6^-$ in benzene at 564 nm, which corresponds to a singlet energy (E^{*1}) of 50.7 kcal/mol.

The electrochemical behavior of the cyanine dyes has been studied extensively.[20] In acetonitrile solution with tetrabutylammonium tetrafluoroborate as the supporting electrolyte, $[Cy]^+$ shows a reversible reduction wave at −1.00 V vs SCE. The reduction generates the cyanine radical, $[Cy]^{\bullet}$, which has an apparent absorption maximum at ~427 nm.

Laser flash photolysis of $Cy^+PF_6^-$ reveals formation of several transient intermediates. An absorption spectrum recorded 60 ps after excitation of $Cy^+PF_6^-$ in benzene solution shows immediate consumption of

$[Cy]^+$ and formation of the excited singlet state of the dye. Most of the absorption intensity of $[Cy]^+$ returns with a lifetime of 250 ± 10 ps. The small residual bleaching of the dye revealed in these spectra is due to the formation of the cyanine mono-*cis* photoisomer. Related experiments conducted on a longer time scale show that the photoisomer decays over several microseconds and that all of the original absorption intensity of $[Cy]^+$ returns on this time scale.

3.3. Alkyltriphenylborates: Chemical and Physical Properties

An issue of primary concern for analysis of the photochemistry of the cyanine borate salts is the oxidation potential of the alkyltriphenylborates. Cyclic voltammetry gives irreversible oxidation waves for each of the borates examined. The peak potentials observed in these experiments vary systematically with the structure of the borate. Those borates bearing alkyl groups capable of forming "stabilized" radicals (i.e., benzyl and 2-naphthylmethyl) have peak potentials lower than those borates that can generate only "unstabilized" radicals (i.e., cyclopropyl and phenyl).

Since electrochemical methods proved unsuitable for measurement of thermodynamically meaningful oxidation potentials of the borates, we determined these values by application of the Weller equation. The rate constant for fluorescence quenching for a series of polycyclic aromatic hydrocarbons by the borates was measured. The reaction was monitored by time-resolved absorption spectroscopy and shown to be electron transfer. Analysis of the fluorescence quenching experiments gives k_{et}, which when substituted into Eqs. 14 and 15, gives the "kinetic" oxidation potentials. As expected for a case where a very fast chemical reaction follows electron transfer, the kinetically determined oxidation potentials (E^{k}_{ox}) for the borates are more positive than are the peak potential values determined by cyclic voltammetry (E^{p}_{ox}).

3.4. Cyanine Borates: Absorption and Ion Pairing in Solution

The electronic absorption spectra of the cyanine borates in benzene solution are independent of the structure of the anion. Thus the oscillator strength of the transition and the energy of $Cy^+(S_1)$ are unaffected by changing the counter-ion from PF_6^- to borates. We conclude, therefore, that excitation of $Cy^+[R–B(Ph_3)]^-$ with visible light generates a locally excited state of $[Cy]^+$, not an exciplex.

Salts dissolved in organic solvents exist as a mixture of freely solvated ions, ion pairs, and higher aggregates whose relative proportion depends on concentration and the details of structure. The state of association plays an important role in controlling the photochemistry of the cyanine borates.

The conductivity of solutions of $Cy^+PF_6^-$ in a series of solvents with different dielectric constants gives values of the dissociation constant for the salt (K_D). The Fuoss equation (Eq. 17), relates the value of K_D to the dielectric constant of the solvent (ε) and the center-to-center distance at contact of the ions in the ion pair (a); the other symbols in the equation have their conventional meanings. A plot of the conductivity data-derived dissociation constants for $Cy^+PF_6^-$ according to the Eq. 17 gives the value of K_D in benzene solution.

$$K_D = [3000/4\pi Na^3]\exp-[e^2/(a\varepsilon k_BT)] \qquad (17)$$

These findings show that the cyanine borates in benzene solution exist essentially exclusively as tight-ion pairs. In solvents with dielectric constants higher than benzene, such as ethyl acetate, the proportion of the cyanine dye associated as ion pairs changes significantly over the concentration range of interest. Tight ion pair formation has important consequences in both the kinetics and thermodynamics of the reactions of electronically excited cyanine borates.

3.5. Laser Flash Photolysis

Irradiation of a benzene solution of $Cy^+[PhCH_2B(Ph)_3]^-$ with an 18 ps pulse instantaneously gives a transient spectrum showing bleaching of the cyanine dye and an absorbance at 430 nm due to $[Cy]^{\bullet}$. There are no meaningful changes in this absorption spectrum for times up to 10,000 ps after the excitation pulse. Related spectral changes are observed following the pulsed irradiation of the other cyanine borates. However, laser flash photolysis of the tetraphenylborate salt shows essentially no formation of $[Cy]^{\bullet}$.

Pulsed irradiation of a benzene solution of $Cy^+[NpCH_2B(Ph)_3]^-$ instantaneously gives the transient absorption spectrum showing formation of the naphthylmethyl radical. This permits estimation of a lower limit for the lifetime of the boranyl radical in this case of 10 ps.

3.6. Intra-Ion Pair Electron Transfer: Estimation of the Free Energy (ΔG_{et})

The free energy for the photoinitiated electron transfer from the borate to the excited cyanine in the ion pair was calculated by application of Eq. 8. The potentials for reduction of $[Cy]^+$ and oxidation of the borates were obtained from experiments carried out in acetonitrile solution. But the photochemical experiments on the cyanine borates were performed in benzene solution where formation of the required ion pairs is complete. Thus calculation of ΔG_{et} requires inclusion of the effect of the solvent change on the electrochemical potentials.

The usual procedure for incorporation of solvent effects on electrochemical potentials is by consideration of the difference in the solvation energies of the ions according to Eq. 18, where $\varepsilon_{(PhH)}$ and $\varepsilon_{(MeCN)}$ are the dielectric constants of benzene and acetonitrile.

$$(E_{ox} - E_{red})_{PhH} = (E_{ox} - E_{red})_{MeCN} -$$

$$e^2/2\,[1/r_- + 1/r_+]][1/\varepsilon_{(PhH)} - 1/\varepsilon_{(MeCN)}] \qquad (18)$$

The change in solvent from acetonitrile to benzene has an enormous effect on the values of the electrochemical potentials of the cyanine and the borates. This is a matter of some concern since the cyanine is certainly not spherical.

The effect of Coulombic stabilization of the cyanine borate ion pair may be calculated from the estimate of K_D extrapolated from the conductivity data by means of the Fuoss equation. This gives a value ~ 0.96 eV in benzene solution. Incorporation of these results into Eq. 8 gives the estimates for ΔG_{et} of the singlet excited cyanine borate ion pairs.

3.7. Fluorescence Spectroscopy Gives k_{et}

The fluorescence efficiency of the cyanine borate salts in benzene solution depends on the identity of the borate anion and provides a direct measure of the rate constant for electron transfer from the borate to the excited state of the cyanine. It is apparent that k_{et} is strongly dependent on the nature of the alkyl group bound to the boron atom of the borate. In general, as the "stability" of the radical formed from carbon–boron bond cleavage increases (i.e., phenyl versus benzyl) the magnitude of k_{et} increases.

3.8. A Mechanism for Intra-Ion Pair Electron Transfer of Cyanine Borates

The experimental observations support postulation of the reaction mechanism shown in Scheme 2. The cyanine and borate exist as an ion pair in benzene solution. This self-association is a prerequisite for the photoinitiated reaction since the lifetime of the excited singlet state of the cyanine is too short to allow an efficient diffusive encounter at experimentally achievable concentrations of the borate. Irradiation of the ion pair with visible light generates the excited singlet state of the cyanine, which can undergo three possible reactions: fluorescence, rotation to a photoisomer, or electron transfer to form the cyanine and boranyl radicals. Once formed, the cyanine–boranyl radical pair may undergo back electron transfer to regenerate the ion pair, or undergo cleavage of a carbon–boron bond to form an alkyl radical [R$^{\bullet}$] and triphenylborane. The [Cy$^{\bullet}$][R$^{\bullet}$] radical pair formed in the latter process might undergo electron transfer to form [Cy^{+}] and an alkyl anion [R^{-}], or the radicals might couple to form an alkylated cyanine [Cy–R], or their diffusion from the initial solvent cage will give free radicals. Analysis of the structural dependence of the rates and efficiencies reveals important details about intra-ion pair electron transfer reactions.

Examination of the fluorescence of the cyanine borate salts and the results of the laser spectroscopy experiments shows that the electron transfer reaction can have two limiting outcomes. For borates capable of generating stabilized alkyl radicals (i.e., benzyl), nearly all of the excited cyanine dye is converted to [Cy]$^{\bullet}$ by the electron transfer during the 18 ps laser pulse. For borates incapable of generating a stabilized radical (i.e., phenyl), there is very little formation of [Cy]$^{\bullet}$. In these cases electron transfer occurs, but back electron transfer (k_{bet}) is much faster than competing processes. For the borates with structures between these two extremes, electron transfer is followed by competition between bond cleavage and back electron transfer.

Rate constants for electron transfer are related to the free energy of the reaction through the classical Marcus equation. A plot of the data for the intra-ion pair electron transfer in the cyanine borates according to this approach is shown in Figure 6. Comparison of the rate constant dependence on the oxidation potentials of the borates determined electrochemically with those measured kinetically shows that the two curves are displaced along the ΔG_{et} axis but that they have approximately the same shape. Since the kinetically determined values are closer to the ther-

Reactants	Rate constant	Products	Process
$[Cy^+][RB(Ph)_3]^-$ Ion Pair	$h\nu$ visible	$[Cy^+]^{*1}[RB(Ph)_3]^-$ Excited ion pair	Excitation
$[Cy^+]^{*1}[RB(Ph)_3]^-$	k_{fl}	$[Cy^+][RB(Ph)_3]^- + h\nu'$	Fluorescence
$[Cy^+]^{*1}[RB(Ph)_3]^-$	k_{rot}	$[iso\text{–}Cy^+][RB(Ph)_3]^-$	Photoisomerization
$[Cy^+]^{*1}[RB(Ph)_3]^-$	k_{et}	$[Cy^\bullet][RB(Ph)_3]^\bullet$	Electron transfer
$[Cy^\bullet][RB(Ph)_3]^-$	k_{bet}	$[Cy^+][RB(Ph)_3]^-$	Back–electron transfer
$[Cy^\bullet][RB(Ph)_3]^\bullet$	k_{-bc} ⇄ k_{bc}	$[Cy^\bullet](R^\bullet)B(Ph)_3$ Solvent cage	Carbon–boron bond cleavage
$[Cy^\bullet](R^\bullet)$		$[Cy^+][R^-]$	
$[Cy^\bullet](R^\bullet)$	k_{CYR}	Cy—R	Alkylation of cyanine
$[Cy^\bullet](R^\bullet)$	k_{diff}	$[Cy^\bullet] + (R^\bullet)$	Cage escape

Scheme 2.

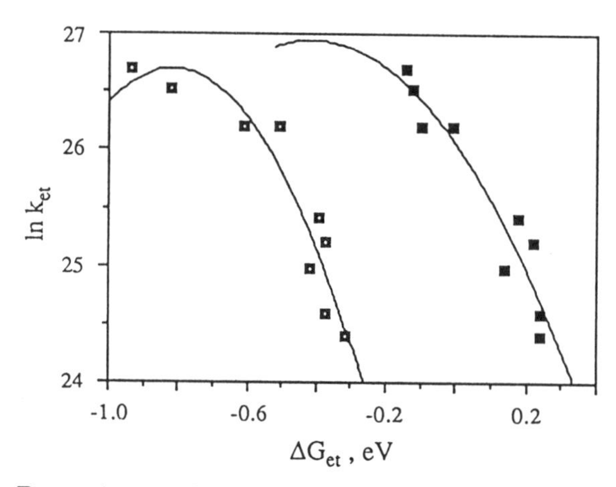

Figure 6. Dependence of the Rate Constant for Intraion Pair Electron Transfer from Borate to Excited Cyanine on ΔG_{et}. The Open Squares Represent Data Obtained from the Peak of the Oxidation Wave for the Borates; the Filled Squares Come from Oxidation Potentials Estimated from the Quenching of the Fluoresence by the Borates.

modynamically meaningful potentials, the latter curve will be used in the analysis of the electron transfer reaction.

Although the range of free energies for the electron transfer reactions covered by the data shown in Figure 6 is narrow and restricted to values near zero, it is possible to reach some tentative conclusions about the functional dependence of k_{et} on ΔG_{et}. The rate constants obtained are far greater than those commonly observed for intermolecular electron transfer reactions since ion pairing eliminates the limitation imposed by diffusion. With these conditions Marcus theory predicts that the electron transfer rates will be controlled by λ. The value of λ_s for a nonpolar solvent such as benzene is readily calculated to be nearly zero. Thus the reorganization energy revealed by our measurements may be associated primarily with λ_v. Analysis of the data shown in Figure 6 according to the Marcus equation (correlation coefficient = 0.95) shows that λ_v is ~0.4 eV. Of course the precise value of λ is dependent on the assumptions made in the calculation of ΔG_{et}. We note, however, that the internal reorganization energy revealed in these experiments is similar to the value recently reported by Gould et al. in related charge recombination reactions.[16] These findings are consistent with our postulate of an electron transfer reaction to generate the cyanine radical and alkyltriphenylboranyl radical pair as intermediates in the photoinitiated reactions of these salts.

According to the mechanism proposed in Scheme 2, the radical pair formed by electron transfer from the borate to the excited cyanine may undergo back electron transfer to regenerate the initial ion pair or it may undergo carbon–boron bond cleavage to form triphenylborane and the alkyl radical in a solvent cage with $[Cy]^{\bullet}$. It is clear from the spectroscopic studies that the efficiency of $[Cy]^{\bullet}$ formation depends on the alkyl group bound to boron. In principle, this effect might be caused by a decrease in the equilibrium constant for carbon–boron bond cleavage in the series as the "stability" of the alkyl radical formed decreases, or it could be due to an increase in the rate constant for back electron transfer (k_{bet}) in this series. Our findings suggest that the first explanation is more reasonable.

According to the Marcus equation, the rate constant for back electron transfer should be determined by the free energy for this reaction, ΔG_{bet}. Values of ΔG_{bet} are calculated from the cyanine reduction potential and the borate oxidation potentials. It is clear for all cases examined that back electron transfer is a very exothermic process and that its exothermicity decreases through the series from $[Ph_4B]^-$ to $[NpCH_2B(Ph)_3]^-$. Since these reactions are run in benzene solution where λ_s is negligible, it is reasonable to expect that these back electron transfers fall in the inverted region where k_{bet} decreases with increasing reaction exothermicity. If this is so, and if the efficiency of radical formation were controlled primarily by k_{bet}, then, contrary to the findings, $[Ph_4B]^-$ should give more radicals than does $[NpCH_2B(Ph)_3]^-$. On this basis we suggest that the radical yield is controlled primarily by the rate of bond cleavage following oxidation of the borate. Nevertheless, the prediction that the rate of the back electron transfer reaction is retarded by inverted region behavior may play an important role in radical formation. If the back electron transfer rates were not somehow slowed, carbon–boron bond cleavage of the boranyl radical might not ever be competitive.

In summary, cyanine borate salts dissolved in nonpolar solvents such as benzene exist exclusively as ion pairs. Irradiation of the ion pair leads to formation of a locally excited singlet state of the cyanine. The rate constant for electron transfer from the borate to the excited cyanine in the ion pair (k_{et}) depends on ΔG_{et}. This latter value was estimated for a series of borates and the relationship between k_{et} and ΔG_{et} appears to give a maximum value consistent with the quadratic law predicted from Marcus theory. The rate of carbon–boron bond cleavage in the boranyl radical depends directly on the stability of the alkyl radical formed. When stabilized alkyl radicals are formed, carbon-boron bond cleavage is faster than the back electron transfer reaction that regenerates the cyanine borate ion pair. This may be due to inhibition of the latter reaction by its large exothermicity in accord with predictions from Marcus theory. Carbon–boron bond cleavage of the boranyl radical in the systems examined is irreversible and leads to cage-escape and the formation of free alkyl and cyanine radicals. The free alkyl radicals formed by the irradiation—electron transfer—bond cleavage sequence may be used in meaningful chemical processes such as the initiation of polymerization.

3.9. Chemically Initiated Electron-Exchange Luminescence (CIEEL)

The annihilation of oppositely charged radical ions releases energy. If the energy released by this process is sufficient, one of the products may be formed in an electronically excited state. This reaction is illustrated in Eq. 19 (conceptually the reverse of Eq. 1). Subsequent emission of light by the excited state formed in the annihilation step accomplishes the chemical formation of light that is known as chemiluminescence.

$$A^{\bullet-} + D^{\bullet+} \rightarrow A + D^* \tag{19}$$

There are several means available for the preparation of the oppositely charged radical ions required for the annihilation reaction. In the most straightforward approach, these reagents are prepared separately by conventional chemical reactions. The chemiexcitation step (the ion annihilation) is then accomplished simply by mixing the two components.[21] A second way to generate the radical ions is electrochemically. In this process the acceptor (A) is first reduced at an electrode, then in the same vessel (often at the same electrode) the donor (D) is oxidized. Diffusion in solution brings the radical ions together in a zone near the electrode surface where they annihilate to form excited states and then emit light. This sequence is called electrogenerated chemiluminescence (ECL); it has been extensively studied.[22]

A third means for generating the oppositely charged radical ions required in Eq. 19 is by a chemical reaction involving a high energy, reactive electron acceptor (an organic peroxide in all of the known examples) and an electron donor. This sequence is illustrated in Scheme 3, it has been designated chemically initiated electron-exchange luminescence (CIEEL).

Two electron transfer reactions are postulated to be required in the CIEEL process. The first is endothermic and represents the rate-determining step of the entire sequence. Since the free energy change for this reaction falls in the normal region we anticipate that the rate of this reaction will increase as the driving force for the process increases. This point will be discussed in more detail below.

The second step in the CIEEL sequence is the fragmentation of the reduced peroxide. Recall from our previous discussion that endothermic electron transfer reactions are "instantaneously" reversible. Thus one might expect that the electron transfer reaction that forms the donor radical cation and the peroxide radical anion will instantaneously reverse and that there will be no net reaction. The saving feature, however, is that the peroxide radical anion fragments rapidly. In this connection, rapidly may be instantaneous, that is, there may be no bound state for the peroxide radical anion. Once the oxygen–oxygen bond of the radical anion has cleaved, back electron transfer to regenerate the neutral donor is endothermic since it would require the oxidation of an oxygen centered radical to a cation or an oxygen centered anion to a radical. Thus the endothermic electron transfer reaction is rendered irreversible by a fast following chemical reaction.

$$\text{ROOR} + \text{D} \underset{k_{-\text{diff}}}{\overset{k_{\text{diff}}}{\rightleftharpoons}} \text{ROOR D} \underset{k_{-\text{et}}}{\overset{k_{\text{et}}}{\rightleftharpoons}} [\text{ROOR}]^{\bullet -}\text{D}^{\bullet +}$$

$$\text{Light} \longleftarrow \text{R}' + \text{D}^{*} \longleftarrow [\text{R}']^{\bullet -}\text{D}^{\bullet +} \longleftarrow [\text{RO}^{-}\,{}^{\bullet}\text{OR}]\text{D}^{\bullet +}$$

Scheme 3.

The next step in the CIEEL sequence outlined in Scheme 3 is the fragmentation of the bond-cleaved peroxide radical anion to generate a new radical anion and, typically, a small fragment molecule such as CO_2. This reaction raises the reducing potential of the radical anion. In one of the more thoroughly studied cases, the starting peroxide is a dioxetanone (Eq. 20). One-electron reduction and oxygen–oxygen bond cleavage generates the ring-opened radical anion. Back electron transfer at this

$$\text{dioxetanone} + \text{D} \xrightarrow{k_{\text{et}}} [\text{dioxetanone}]^{\bullet -} + \text{D}^{\bullet +} \xrightarrow{-CO_2} (CH_3)_2C{=}O^{\bullet -}\ \text{D}^{\bullet +} \longrightarrow \text{D}^{*} \qquad (20)$$

stage of the reaction is endothermic. In the next step the carbon–carbon bond fragments to give CO_2 and the acetone radical anion. The acetone radical anion is a much more powerful reducing agent than is the ring-cleaved peroxide radical anion. Thus, in a sense, one can view the energy flow in the reaction as from chemical bonds to reducing power. That is, the exothermic peroxide decomposition fuels the separation of the electron (on the acetone radical anion) and the hole (the donor radical cation). If these bond-cleaving reactions occur rapidly, then the donor radical cation and the peroxide-derived radical anion will be formed in the same solvent cage.

In the penultimate step of the CIEEL sequence, the oppositely charged radical ions annihilate. If this annihilation is sufficiently exothermic, then an electronically excited state, typically of the donor, may be formed. Emission of light from this excited state completes the chemically initiated electron-exchange luminescence sequence.

From the point of view of this chapter, the two key steps in the CIEEL sequence are the initial endothermic electron transfer and oxygen–oxygen bond cleavage and the back electron transfer (ion pair annihilation) that generates the electronically excited state. We will examine these two steps in some detail.

3.10. CIEEL from Dimethyldioxetanone: Endothermic Electron Transfer

The chemiluminsescent reactions of dioxetanones are of particular interest because of their demonstrated involvement in the bioluminescence of several organisms including that of the well-known North American firefly.[23] Dimethyldioxetanone was prepared first by Adam and co-workers, who showed that its thermolysis generates acetone and CO_2.[24] Some of the acetone formed in this reaction is electronically excited (predominantly in its triplet state) but the yield is low (1.6%).[25]

We found that the decomposition of dimethyldioxetanone was accelerated by the addition of electron donors to the reaction solution.[26] Moreover, this rate acceleration was accompanied by an increase in the yield of excited singlet states, which were found to be predominantly localized on the donor. Analysis of the absorption spectra of the donors in the presence of the peroxide showed no change except for the special case of some metalloporphyrins.[27] Thus we conclude that normally there is no strong ground state interaction between the donor and the peroxide.

Electron transfer reactions that occur in the normal energy region are expected to show a nearly linear free energy relationship. That is, for large positive values of $\Delta G^{\ddagger}$, a plot of log k_{et} against ΔG_{et} should give a straight line. Analysis of the Marcus equation for reactions of this sort leads to the prediction that the slope of this line should be 0.5. For the case of dimethyldioxetanone, such a plot is linear but its slope is only ~0.3. This deviation from the predicted value led to some concern that the rate-determining step in the reaction is only a "partial" electron transfer, but this was shown to be not a required interpretation.[28,29]

A key feature to consider in the endothermic electron transfer reaction is the relationship between the stretching of the oxygen–oxygen bond and the reaction coordinate for electron transfer. We presume that an encounter complex is formed between the donor and the peroxide. This complex does not strongly perturb the electronic character of the system; the oxygen–oxygen bond, for example, has its normal length in the complex. Transfer of an electron from the donor to the peroxide at the geometry of the encounter complex is endothermic and does not occur

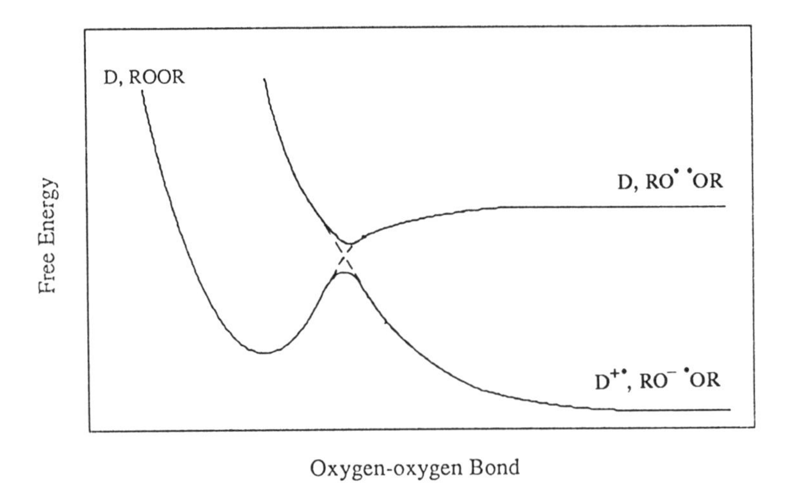

Figure 7. Dissociative Electron Transfer from an Electron Donor (D) to a Peroxide. There is No, or Very Little, Oxygen–Oxygen Bond Energy in the Peroxide Radical Anion.

spontaneously. Consider the consequence, however, of stretching the oxygen–oxygen bond of the peroxide in the encounter complex (Figure 7). In the absence of a donor at some distance the oxygen-oxygen bond will be broken and a diradical will be formed. However, in the presence of a good donor, a spontaneous (exothermic) electron transfer to the bond-cleaved peroxide can occur to form the donor radical cation and the cleaved peroxide radical anion. These facts ensure that at some oxygen–oxygen bond distance, the transition structure for the reaction, electron transfer will occur followed by a further barrierless increase of the oxygen–oxygen bond distance. This scenario identifies oxygen–oxygen bond stretching as the key internal mode (λ_i) for activation of the endothermic electron transfer. The more powerful the electron donor, the less stretching of the oxygen–oxygen bond is required to reach the transition structure. If the potential energy surface for stretching the oxygen–oxygen bond is anharmonic, then there is no reason to expect that the slope of the plot of log k_{et} against ΔG_{et} will be 0.5.

3.11. CIEEL from Dimethyldioxetanone: Exothermic Back Electron Transfer

The second step in the electron-exchange sequence is the annihilation of the oppositely charged radical ions. Conceptually, this can be viewed as a back electron transfer reaction with the unusual complication that the reducing potential of the electron has been raised considerably before it is transferred back. From the point of view of this chapter, the unusual aspect of this reaction is that it forms electronically excited states. This feature has been taken as some indication for the existence of the Marcus inverted region.[30]

Consider two limiting outcomes for the ion annihilation reaction: generation of excited state or ground state products. If we presume that we have chosen the reagents so that electronically excited state formation is energetically possible, then the free energy change for the back electron transfer (ΔG_{bet}) must be large and negative. Since these reactions can be carried out in nonpolar solvents where λ_s is small, it is likely that the back electron transfer to form ground state products falls in the inverted range. When the product of the annihilation is an electronically excited state, the excitation energy ΔE^* must be subtracted from ΔG_{bet}. Typically ΔE^* is in the range of 50–75 kcal/mol and thus the ion annihilation is not nearly as exothermic when an excited state is produced.

In some reactions, ion annihilation to form excited products occurs with very high efficiency. The rate restriction of inverted region behavior rationalizes this outcome. The back electron transfer to form ground state products is slowed because of its high exothermicity, the back electron transfer to form excited states is not as strongly retarded because it is less exothermic.

4. CONCLUSIONS

In this chapter we examined the parameters that determine the driving force for electron transfer reactions and briefly explored the role of Marcus theory in predicting the rate constant for electron transfer from the knowledge of the driving force. These general considerations were probed in two specific contexts: intra-ion pair electron transfer of cyanine borates and chemically inititated electron-exchange luminescence. In both specific examples, the data indicate that an initial electron transfer is followed by a rapid and irreversible cleavage of a bond. It is the bond cleavage step that ensures a useful reaction.

ACKNOWLEDGMENT

This work was supported by grants from the National Science Foundation and by Mead Imaging Corporation, for which we are grateful.

REFERENCES

1. Eberson, L. *Electron Transfer Reactions in Organic Chemistry.* Springer-Verlag, Berlin, 1987. Kavarnos, G. J.; Turro, N. J. *Chem. Rev.* **1986**, *86*, 401. Mattay, J. *J. Synth. Org. Chem.* **1989**, 233. Chibisov, A. K. *Russ. Chem. Rev.* **1981**, 615. Kosower, E. M.; Huppert, D. *Annu. Rev. Phys. Chem.* **1986**, *37*, 127. Kuzmin, M. G.; Soboleva, V. *Prog. Reaction Kinet.* **1986**, *14*, 157. Pac, C.; Ishitani, O. *Photochem. Photobiol.* **1986**, *48*, 767. Reimers, J. R.; Hush, N. S. *Chem. Phys.* **1989**, *134*, 323.
2. Chibisov, K. *Russ. Chem. Rev. (Engl. Transl.)* **1981**, *50*, 1169. Born, M. *Z. Phys.* **1920**, *1*, 45. Weller, A. *Z. Phys. Chem. Neue Folge* **1982**, *133*, 93.
3. Atkins, P. W.; MacDermott, A. J. *J. Chem. Educ.* **1982**, *59*, 359.
4. Suppan, P. *Chimia* **1988**, *42*, 320. Suppan, P. *J. Chem. Soc. Faraday Trans. I* **1986**, *82*, 509.
5. Glasstone, S.; Laidler, K. J.; Eyring, H. *Theory of Rate Processes.* Academic Press, New York, 1948, p. 387.
6. Verhoven, J. W. *Pure Appl. Chem.* **1986**, *58*, 1285. Heitele, H.; Finckh, P.; Weeren, S.; Pollinger, F.; Michel-Beyerle, M. E. *J. Phys. Chem.* **1989**, *93*, 5173. Schmidt, J. A.; Liu, J.-Y.; Bolton, J. R.; Archer, M. D.; Gadzekpo, V. P. Y. *J. Chem. Soc. Faraday Trans. I* **1989**, 1027.
7. Fukui, K. *Acc. Chem. Res.* **1971**, *4*, 57. Turro, N. J. *Modern Molecular Photochemistry.* Benjamin, California, 1978, p. 201.
8. Sjostrom, M.; Wold, S. *Acta. Chem. Scand.* **1981**, *B35*, 537. Kamlet, M. J.; Taft, R. W. *Acta Chem. Scand.* **1985**, *B39*, 611.
9. Marcus, R. A. *Annu. Rev. Phys. Chem.* **1964**, *15*, 155.
10. Kestner, N. R.; Logan, J.; Jortner, J. *J. Phys. Chem.* **1982**, *86*, 622.
11. Rehm, D.; Weller, A. *Ber. Bunsenges. Phys. Chem.* **1969**, *73*, 834.
12. Closs, G. L.; Miller, J. R. *Science* **1988**, *240*, 440.
13. Efrima, S.; Bixon, M. *Chem. Phys. Lett.* **1974**, *25*, 34.
14. Kakitani, T.; Mataga, N. *Chem. Phys. Lett.* **1986**, *124*, 437. Kakitani, T.; Mataga, N. *J. Phys. Chem.* **1986**, *90*, 993.
15. Miller, J. R.; Calcaterra, L. T.; Closs, G. L.; *J. Am. Chem. Soc.* **1984**, *106*, 3047.
16. Gould, I. R.; Moser, J. E.; Ege, D.; Moody, R.; Armitage, B.; Farid, S. *J. Imag. Sci.* **1989**, *33*, 44.
17. Marcus, R. A. *J. Phys. Chem.* **1989**, *93*, 3078.
18. Tachiya, M. *Chem. Phys. Lett.* **1989**, *159*, 505.

19. Beitz, J. V.; Miller, J. R. *J. Chem. Phys.* **1979**, *71*, 4579.
20. Lenhard, J. *J. Imaging Sci.* **1986**, *30*, 27.
21. Schuster, G. B.; Schmidt, S. P. *Adv. Phys. Org. Chem.* **1982**, *18*, 187.
22. Faulkner, L. R. *Int Rev. Sci.: Phys. Chem. Ser. Two* **1976**, *9*, 213.
23. Wannland, J.; DeLuca, M.; Stemple, K.; Boyer, P. D. *Biochem. Biophys. Res. Commun.* **1978**, *81*, 987.
24. Adam, W.; Liu, J.-C.; *J. Am. Chem. Soc.* **1972**, *94*, 2894.
25. Schmidt, S. P.; Schuster, G. B. *J. Am. Chem. Soc.* **1980**, *102*, 306.
26. Schmidt, S. P.; Schuster, G. B. *J. Am. Chem. Soc.* **1978**, *100*, 1966.
27. Schmidt, S. P.; Schuster, G. B. *J. Am. Chem. Soc.* **1980**, *102*, 7100.
28. Walling, C. *J. Am. Chem. Soc.* **1980**, *102*, 6855.
29. Scandola, F.; Balzani, V.; Schuster, G. B. *J. Am. Chem. Soc.* **1981**, *103*, 2519.
30. Marcus, R. A. *J. Chem. Phys.* **1965**, *43*, 2654.

JAI PRESS

Advances in Cycloaddition

Edited by **Dennis P. Curran,** *Department of Chemistry, University of Pittsburgh*

"It is the intention of this volume to begin a serial coverage of the broad areas of cycloaddition chemistry. Cycloaddition reactions are among the most powerful reactions available to the organic chemist. The ability to simultaneously form and break several bonds, with a wide variety of atomic substitution patterns, and often with a high degree of stereocontrol, has made cycloaddition reactions the subject of intense study. The productive interplay between theory and experiment has resulted in sophisticated models which often allow one to predict reactivity, regioselectivity, and steroselectivity for given cycloaddition partners."

—*From the Preface to Volume 1*

REVIEW: This volume is highly recommended to all those who want to stay abreast of developments in the mechanisms and synthetic applications of 1,3-dipolar cycloaddition reactions. The writers have realized a good balance between the summary of achievements and the reporting of gaps in understanding or remaining synthetic challenges. The articles are well written, they are amply illustrated with equations or schemes, and they cover the literature into or through 1986.

- *Journal of the American Chemical Society*

Volume 1, 1988, 208 pp. $78.50
ISBN 0-89232-861-4

CONTENTS: List of Contributors. Introduction to the Series: An Editor's Foreword, *Albert Padwa, Emory University.* **Preface,** *Dennis P. Curran.* **Steric Course and Mechanism of 1,3-Dipolar Cycloadditions,** *Rolf Huisgen, Institut fur Organische Chemie der Universitat Munchen.* **Nonstabilized Azomethine Ylides,** *Edwin Vedejs, University of Wisconsin.* **Molecular Rearrangements Occurring from Products of Intramolecular 1,3 Dipolar Cycloadditions: Synthetic and Mechanistic Aspects,** *Arthur G. Schultz, Rensselaer Polytechnic University.* **Dipolar Cycloadditions of Nitrones with Vinyl Ethers and Silane Derivatives,** *Philip DeShong, University of Maryland, Stephen W. Lander, Jr., United States Air Force Academy, Joseph M. Leginus, Sandoz, Inc. and C. Michael Dickson, Smith-Klein and French Laboratories.* **The Cycloaddition Approach to ^{5}-Hydroxy Carbonyls: An Emerging Alternative to the Aldol Strategy,** *Dennis P. Curran, University of Pittsburgh.*

Volume 2, 1990, 220 pp. $78.50
ISBN 0-89232-951-3

CONTENTS: List of Contributors. Introduction to the Series: An Editor's Foreword, *Albert Padwa.* **Preface,** *Dennis P. Curran.* **Intramolecular 1,3-Dipolar Cycloaddition Chemistry,** *Albert Padwa and Allen M. Schoffstall, Emory University.* **Stereochemical and Synthetic Studies of the Intramolecular Diel-Alder Reaction,** *William R. Roush, Indiana University, Bloomington.* **Thermal Reaction of Cyclopropenone Ketals, Key Mechanistic Features, Scope and Application of the Cycloaddition Reactions of Cyclopropenone Ketals and - Delocalized Singlet Vinyl Carbenes; Three Carbon I,I-/1,3-Dipoles,** *Dale L. Boger, Purdue University, and Christine E. Brotherton-Pleiss, Syntex Research Institute of Bioinorganic Chemiistry, Palo Alto.*

Volume 3, In preparation, Fall 1991
ISBN 1-55938-319-4 Approx. $78.50

J A I P R E S S

JAI PRESS INC.
55 Old Post Road - No. 2
P.O. Box 1678
Greenwich, Connecticut 06836-1678
Tel: 203-661-7602

JAI PRESS

Advances in Molecular Vibrations and Collision Dynamics

Edited by **Joel M. Bowman,** *Department of Chemistry, Emory University*

Volume 1 - Part A, In preparation, Summer 1991
ISBN 1-55938-294-5 Approx. $78.50
Set ISBN 1-55938-293-7 Set Price: Approx. $157.00

CONTENTS: An Introduction to the Dynamics of van der Waals Molecules, *Jeremy M. Hutson, University of Durham.* **The Nature and Decay of Metastable Vibrations: Classical and Quantum Studies of van der Waals Molecules,** *Stephen K. Gray, Northern Illinois University.* **Optothermal Vibrational Spectroscopy of Molecular Complexes,** *R.E. Miller, University of North Carolina.* **High Resolution IR Laser Driven Vibrational Dynamics in Supersonic Jets: Weakly Bound Complexes and Intramolecular Energy Flow,** *Andrew McIlroy and David J. Nesbitt.* **Three Dimensional Quantum Scattering Studies of Transition State Resonances: Results for O + HCl OH + Cl,** *Hiroyasu Koizumi, Northwestern University and George C. Schatz, Argonne National Laboratory.* **Negative Ion Photodetachment as a Probe of the Transition State Region: The + HI Reaction,** *Daniel M. Neumark, University of California, Berkeley.* **Rovibrational Spectroscopy of Transition States,** *James J. Valentini, University of California, Irvine.* **Optimal Control of Molecular Motion: Making Molecules Dance,** *Herschel Rabitz and Shenghua Shi, Princeton University.* **Static Self Consistent Field Methods for Anharmonic Problems: An Update,** *Mark A. Ratner, Northwestern University, Robert B. Gerber, The Hebrew University and University of California, Irvine, Thomas R. Horn, Northwestern University and The Hebrew University, and Carl J. Williams, Northwestern University.* **Perturbative Studies of the Vibrations of Polyatomic Molecules Using Curvilinear Coordinates,** *Anne B. McCoy and Edwin L. Sibert III, University of Wisconsin-Madison.*

Volume 1 - Part B, In preparation, Summer 1991
ISBN 1-55938-295-3 Approx. $78.50
Set ISBN 1-55938-293-7 Set Price: Approx. $157.00

CONTENTS: Preface. Classical Dynamics and the Nature of Highly Excited Vibrational Eigenstates, *Michael J. Davis, Argonne National Laboratory, Craig C. Martens, University of California, Irvine, Robert G. Littlejohn and J.S. Pehling, University of California, Berkeley.* **Semiclassical Mechanisms of Bound**

and Unbound States of Atoms and Molecules, *David Farrelly, University of California, Los Angeles.* **Dissociation of Overtone-Excited Hydrogen Peroxide Near Threshold: A Quasiclassical Trajectory Study,** *Yuhua Guan, Brookhaven National Laboratory, Turgay Uzer, Brian D. Macdonald, Georgia Institute of Techhology, and Donald L. Thompson, Oklahoma State University.* **L2 Approaches to the Calculation of Resonances in Polyatomic Molecules,** *Bela Gazdy and Joel M. Bowman, Emory University.* **Analytic MBPT(2) Energy Derivaties: A Powerful Tool for the Interpretation and Prediction of Vibrational Sepctra for Unusual Molecules,** *Rodney J. Bartlett, John F. Stanton and John D. Watts, University of Florida, Gainesville.* **Spectro-A Program for the Derivation of Spectroscopic Constants from Provided Quartic Force Fields and Cubic Dipole Fields,** *Jeffrey F. Gaw, Monsanto Company, Andrew Willetts, William H. Green and Nicholas C. Handy, University Chemical Laboratory, England.* **Photoinitiated Reactions in Weakly-Bonded Complexes: Entrance Channel Specificity,** *Y. Chen, G. Hoffmann, S.K. Shin, D. Oh, S. Sharpe, Y.P. Zeng, R.A. Beaudet and C. Wittig, Universtiy of Southern California, Los Angeles.* **Photodissociation Dynamics of the Nitrosyl Halides: The Influence of Parent Vibrations,** *Charles Qian and Hanna Reisler, University of Southern California, Los Angeles.* **Gas-Phase Metal Ion Solvation: Spectroscopy and Simulation,** *James M. Lisy, University of Illinois at Urbana-Champaign.* **The Chemical and Physical Properties of Vibration-Rotation Eigenstates of $H_2CO(SO)$ at 28,300 CM^{-1}** *Hai-Lung Dai, University of Pennsylvania.*

J A I

P R E S S

of MM2, *Joseph J. Gajewski, Kevin E. Gilbert, Indiana University and John McKelvey, Eastman Kodak Company.* **Empirical Derivation of Molecular Mechanics Parameter Sets: Application to □-Lactams,** *Kathleen A. Durkin, Michael J. Sherrod, and Dennis Liotta, Emory University.* **Application of Molecluar Mechanics to the Study of Drug-Membrane Interactions: The Role of Molecular Conformation in the Passive Membrane Permeability of Zidovudine (AZT),** *George R. Painter, John P. Shockcor, and C. Webster Andrews, Burroughs Wellcome Company.*

Volume 3, In preparation, Spring l992
ISBN 1-55938-326-7 Approx. $78.50

JAI PRESS INC.
55 Old Post Road - No. 2
P.O. Box 1678
Greenwich, Connecticut 06836-1678
Tel: 203-661-7602

J A I P R E S S

Advances in Carbocation Chemistry

Edited by **Xavier Creary,** *Department of Chemistry, University of Notre Dame*

Volume 1, 1989, 253 pp. $78.50
ISBN: 0-89232-860-2

"This series will provide a detailed account of some of the recent work in the carbocation area. While each chapter gives a historical perspective, the emphasis of each chapter will be from the viewpoint of the individual authors. The contributions from the authors' laboratories are given primary consideration and no attempt has been made to present a balanced view in controversial areas."

- From the Preface to Volume 1

CONTENTS: List of Contributors. Introduction to Series: An Editor's Foreword, *Albert Padwa.* **Preface,** *Xavier Creary.* **Flourine Substituted Carbocations,** *Annette D. Allen and Thomas T. Tidwell, University of Toronto.* **[5]-Carbonyl Cations,** *Xavier Creary, University of Notre Dame, Alan C. Hopkinson and Edward Lee-Ruff, York University.* **The 2-Norbornyl Carbonium Ion Stabilizing Conditions: An Assessment of Structural Probes,** *George M. Kramer and Charles G. Scouten, Corporate Research Laboratories of Exxon Research & Engineering Company.* **Simple Relationships Between Carbocation Lifetime and the Mechanism for Nucleophilic Substitution at Saturated Carbon,** *John P. Richard, University of Kentucky.* **Generation and Ion-Pair Structures of Unstable Carbocation Intermediates in Solvolytic Reactions,** *Kunio Okamoto, Meisei Chemical Works, Inc., Tokyo.* **Carbocations Destabilized by Electron-Withdrawing Groups: Applications in Organic Synthesis,** *Micheline Charpentier-Morize and Daniele Bonnet-Delpon, Centre National de la Recherche Scientifique, Thiais, France.*

Volume 2, In preparation, Summer 1991
ISBN 0-89232-952-1 Approx. $78.50

JAI PRESS INC.
55 Old Post Road - No. 2
P.O. Box 1678
Greenwich, Connecticut 06836-1678
Tel: 203-661-7602